Marvels

Or, A Year's Microscopic Recreations Among the Polyps, Infusoria, Rotifers, Water-bears and Polyzoa

Henry James Slack

Alpha Editions

This edition published in 2022

ISBN : 9789356909311

Design and Setting By
Alpha Editions
www.alphaedis.com
Email - info@alphaedis.com

Contents

INTRODUCTION.

As this little book is intended to be no more than an introduction to an agreeable branch of microscopical study, it is to be hoped it will not require a formal preface; but a few words may be convenient to indicate its scope and purpose.

The common experience of all microscopists confirms the assertion made by Dr. Goring, that the most fascinating objects are living creatures of sufficient dimensions to be easily understood with moderate magnification; and in no way can objects of this description be so readily obtained, as by devoting an occasional hour to the examination of the little ponds which are accessible from almost any situation. A complete volume of pond lore would not only be a bulky book—much bigger than the aldermanic tomes which it is the fashion to call "Manuals," although the great stone fists in the British Museum would be required to grasp them comfortably,—but its composition would overtask all the philosophers of our day. In good truth, a tea-spoonful of water from a prolific locality often contains a variety of living forms, every one of which demands a profound and patient study, if we would know but a few things concerning it.

To man, then, is a vast and a minute. Our minds ache at the contemplation of astronomical immensities, and we are apt to see the boundless only in prodigious masses, countless numbers, and immeasurable spaces. The Creative Mind knows no such limitations; and the microscope shows us that, whether the field of nature's operation be what to our apprehension is great or small, there is no limit to the exhibition of marvellous skill. If the "undevout astronomer" be "mad," the undevout microscopist must be still more so, for if the matter be judged by human sense, the skill is greater as the operation is more minute; and not the sun itself, nor the central orb round which he revolves, with all his attendant worlds, can furnish sublimer objects of contemplation, than the miraculous assemblage of forces which make up the life of the smallest creature that the microscope reveals.

There is an irresistible charm in the effort to trace *beginnings* in nature. We know that we can never succeed; that each discovery, which conducts back towards some elementary law or principle, only indicates how much still lies behind it: but the geologist nevertheless loves to search out the first or oldest traces of life upon our globe; and so the microscopist delights to view the simplest exhibitions of structures and faculties, which reach their completion in the frame and mind of man. That one great plan runs through the whole universe is now an universally accepted truth, and when applied to physiology and natural history, it leads to most important results.

The researches of recent philosophers have shown us that nature cannot be understood by studying the parts of animals with reference merely to their utility in the economy of the creature to which they belong. We do, indeed, find an admirable correspondence between structures and the services they perform; but every object in creation, and every part of it, is in harmonious relation to some grand design, and exhibits a conformity to some general mode of operation, or some general disposition and direction of forces, which secures the existence of the individual or the species, and at the same time works out the most majestic schemes. Microscopic researches, such as are within the reach of millions, offer many of the most beautiful illustrations of these truths; and although the following pages are confined to such objects as are easily obtainable from ponds, and relate almost exclusively to the Infusoria, the Rotifers, the Polyps, and the Polyzoa, it is hoped that they will assist in associating a few of the highly suggestive reasonings of science, with one of the most pleasurable recreations that human ingenuity has devised.

After a preliminary chapter, which is intended to assist the young microscopist in some technical matters, that could not be conveniently introduced into the text, the observations are distributed in chapters, corresponding with the twelve calendar months. This arrangement was suggested by the author's diary of operations for the year 1860, and although it by no means follows that the months in which particular creatures were then discovered, will be those in which they will be most readily found in other years, it was thought advantageous to give a real account of an actual period of microscopic work, and also that the plan would facilitate a departure from the dry manner of a technical treatise. The index will enable any one to use the book for the purpose of reference, and it will be observed that the first chapter in which any member of a group of creatures is introduced, is that in which a general description of the class is given. The illustrations are taken from drawings made by the wife of the author from the actual objects, with the exception of a few instances, in which the authority is acknowledged. The sketches were made *especially for beginners*, and the rule followed, was not to introduce any details that could not be seen at one focus, and with the simplest means: more elaborate representations, though of the highest value to advanced students, are bewildering at the commencement.

The ponds referred to are all either close to, or within a moderate distance of, London;[1] but similar objects will in all probability be obtained from any ponds similarly situated, and the descriptions and directions given for the capture of the minute prey will be found generally applicable. Care has been taken throughout to explain the most convenient methods of examining the objects, and although verbal descriptions are poor substitutes for the teachings of experience, it is hoped that those here given will remove some difficulties from a pursuit that no intelligent person can enter upon without pleasure, or consent to abandon when its elementary difficulties have been mastered, and the boundless fields of discovery are opened to view. Let not the novice be startled at the word "discovery." It is true that few are likely to arrive at new principles or facts which will inscribe their names upon the roll of fame; but no one of ordinary powers can look at living objects with any considerable perseverance, without seeing much that has never been recorded, and which is nevertheless worthy of note; and when the mind, by its own exertions, first arrives at a knowledge of new truth, an emotion is felt akin to that which more than recompenses the profoundest philosopher for all his toil.

[1] Many are now (1871) destroyed by the progress of building.

CHAPTER I.

PLAIN HINTS ON MICROSCOPES AND THEIR MANAGEMENT.

Powers that are most serviceable—Estimated by focal length—Length of body of microscope and its effects—Popular errors about great magnification—Modes of stating magnified power—Use of an "Erector"—Power of various objectives with different eye-pieces—Examination of surface markings—Methods of illumination—Direct and oblique light—Stage aperture—Dark ground illumination—Mode of softening light—Microscope lamps—Care of the eyes.

THE microscope is rapidly becoming the companion of every intelligent family that can afford its purchase, and, thanks to the skill of our opticians, instruments which can be made to answer the majority of purposes may be purchased for three or four guineas, while even those whose price is counted in shillings are by no means to be despised. The most eminent English makers, Wales, and Tolles, in America, and Hartnack, in Paris, occupy the first rank, while the average productions of respectable houses exhibit so high a degree of excellence as to make comparisons invidious. We shall not, therefore, indulge in the praises of particular firms, but simply recommend any reader entering upon microscopic study to procure an achromatic instrument, if it can be afforded, and having at least two powers, one with a focus of an inch or two thirds of an inch, and the other of half or a quarter. Cheap microscopes have usually only one eye-piece, those of a better class have two, and the best are furnished with three, or even more.

The magnifying power of a compound microscope depends upon the focal length of the object-glass (or glass nearest the object), upon the length of the tube, and the power of the eye-piece. With regard to object-glasses, those of shortest focal length have the highest powers, and the longest eye-pieces have the lowest powers. The body of a microscope, or principal tube of which it is composed, is, in the best instruments, about nine inches long, and a draw tube, capable of being extended six inches more, is frequently useful. From simple optical principles, the longer the tube the higher the power obtained with the same object-glass; but only object-glasses of very perfect construction will bear the enlargement of their own imperfections, which results from the use of long tubes; and consequently for cheap instruments the opticians often limit the length of the tube, to suit the capacity of the object-glasses they can afford to give for the money. Such microscopes may be good enough for the generality of purposes, but they do not, with glasses of given focal length, afford the same magnifying power as is done by instruments of better construction. The best and most expensive glasses will not only bear long tubes, but also eye-pieces of high power, without any practical diminution of the accuracy of their operation, and this is a great convenience in natural history investigations. To obtain it, however, requires such perfection of workmanship as to be incompatible with cheapness. An experienced operator will not be satisfied without having an object-glass at least as high as a quarter, that will bear a deep eye-piece, but beginners are seldom successful with a higher power than one of half-inch focus, or thereabouts, and before trying this, they should familiarise themselves with the use of one with an inch focus.

It is a popular error to suppose that enormous magnification is always an advantage, and that a microscope is valuable because it makes a flea look as big as a cat or a camel. The writer has often smiled at the exclamations of casual visitors, who have been pleased with his microscopic efforts to entertain them. "Dear me, what a wonderful instrument; it must be immensely powerful;" and so forth. These ejaculations have often followed the use of a low power, and their authors have been astonished at receiving the explanation that the best microscope is that which will show the most with the least magnification, and that accuracy of definition, not mere increase of bulk, is the great thing needful.

Scientific men always compute the apparent enlargement of the object by *one* dimension only. Thus, supposing an object one hundredth of an inch square were magnified so as to appear one inch square, it would, in scientific parlance, be magnified to "one hundred diameters," or one hundred linear; and the figures 100 would be appended to any drawing which might be made from it. It is, however, obvious that the length is magnified as well as the breadth; and hence the magnification of the whole surface, in the instance specified, would be one hundred times one hundred, or ten thousand: and this is the way in which magnification is popularly stated. A few moments' consideration will show that the scientific method is that which most readily affords information. Any one can instantly comprehend the fact of an object being made to look ten times its real length; but if told that it is magnified a hundred times, he does not know what this really means, until he has gone through the process of finding the square root of a hundred, and learnt that a hundredfold magnification means a tenfold magnification of each superficial dimension. If told, for example, that a hair is magnified six hundred diameters, the knowledge is at once conveyed that it looks six hundred times as broad as it is; but a statement that the same hair is magnified three hundred and sixty thousand times, only excites a gasping sensation of wonder, until it is ascertained by calculation that the big figures only mean what the little figures express. In these pages the scientific plan will always be followed.

If expense is not an object, a binocular instrument should be purchased, and it is well to be provided with an object-glass as low as three or even four inches focus, which will allow the whole of objects having the diameter of half an inch or more to be seen at once. Such a low power is exceedingly well adapted for the examination of living insects, or of the exquisite preparations of entire insects, which can now be had of all opticians. Microscopes which have a draw tube can be furnished with an *erector*, an instrument so called because it erects the images, which the microscope has turned upside down, through the crossing of the rays. This is very convenient for making dissections under the instrument; and it also gives us the means of reducing the magnifying power of an object-glass, and thus obtaining a larger field. The erector is affixed to the end of the draw tube, and by pulling it out, or thrusting it in, the rays from the object-glass are intercepted at different distances, and various degrees of power obtained.

A binocular microscope is most useful with low powers from two thirds upwards. A new form, devised by Mr. Stephenson, acts as an erector, and is very valuable for dissections. It works with high powers.

Beginners will be glad to know how to obtain the magnifying power which different objects require, and it may be stated that, with a full-sized microscope, a two-inch object-glass magnifies about twenty-five diameters with the lowest eye-piece; a one-inch object-glass, or two thirds, from fifty to sixty diameters; a half-inch about one hundred; a quarter-inch about two hundred. The use of deeper eye-pieces adds very considerably to the power, but in proportions which differ with different makers. One instrument used by the writer has three eye-pieces, giving with a two thirds object-glass powers of sixty one hundred and five, and one hundred and eighty respectively; and with a fifth two hundred and forty, four hundred and thirty, and seven hundred and twenty, which can be augmented by the use of a draw tube.

It has been well observed that the illumination of objects is quite as important as the glasses that are employed, and the most experienced microscopists have never done learning in this matter. Most microscopes are furnished with two mirrors beneath the stage, one plane and one concave. The first will throw a few parallel rays through any transparent object properly placed, and the latter causes a number of rays to converge, producing a more powerful effect. The first is usually used in daylight, when the instrument is near a window (one with a north aspect, out of direct sunlight, being the best); and the second is often useful when the source of illumination is a candle or a lamp. By varying the angle of the mirror the light is thrown through the object more or less obliquely, and its quantity should never be sufficient to pain the eye. Few objects are seen to the best advantage with a *large* pencil of perfectly direct light, and the beginner should practise till the amount of inclination is obtained which produces the best effect.

It is advisable that the hole in the stage of the microscope should be large—at least an inch and a half each way—so that the entrance of oblique rays is not obstructed, and it is desirable that the mirror, in addition to sliding up and down, should have an arm by which it can be thrown completely out of the perpendicular plane of the body of the instrument. This enables such oblique rays to be employed as to give a dark field, all the light which reaches the eye being *refracted* by the object through which it is sent. The opticians sell special pieces of apparatus for this purpose, but though they are very useful, they do not render it less desirable to have the mirror mounted as described.

Most microscopes are furnished with a revolving diaphragm, with three holes, of different sizes, to diminish the quantity of light that is admitted to the object. This instrument is of some use, and offers a ready means of obtaining a very soft agreeable light for transparent objects, viewed with low powers. For this purpose cut a circular disk of India or tissue paper, rather larger than the biggest aperture; scrape a few little pieces of spermaceti, and place them upon it, then put the whole on a piece of writing-paper, and hold it a few inches above the flame of a candle, moving it gently. If this is dexterously done, the spermaceti will be melted without singeing the paper, and when it is cold the disk will be found transparent. Place it over the hole in the diaphragm, send the light through it, and the result will be a very soft agreeable effect, well suited for many purposes, such as viewing sections of wood, insects mounted whole, after being rendered transparent, many small water creatures, etc. Another mode of accomplishing this purpose is to place a similarly prepared disk of paper on the flat side of a bull's-eye lens, and transmit the light of a lamp through it. This plan may be used with higher powers, and the white opaque light it gives may be directed at any angle by means of the mirror beneath the stage.

An ordinary lamp may be made to answer for microscopic use, but one of the small paraffine lamps now sold everywhere for eighteen-pence is singularly convenient. It is high enough for many purposes, and can easily be raised by one or more blocks. A paraffine lamp on a sliding stand is still more handy, and all the better for a hole with a glass stopper, through which the fluid can be poured.

Many people fancy that the eyes are injured by continual use of the microscope, but this is far from being the case if reasonable precautions are taken. The instrument should be inclined at a proper angle, all excess of light avoided, and the object brought into focus before it is steadily looked at. Most people solemnly shut one eye before commencing a microscopic examination; this is a practical and physiological mistake. Nature meant both eyes to be open, and usually resents a prolonged violation of her intentions in this matter. It requires but a little practice to keep both eyes open, and only pay attention to what is seen by that devoted to the microscope. The acquisition of this habit is facilitated, and other advantages gained, by a screen to keep out extraneous light. For this purpose take a piece of thin cardboard about nine inches square, and cut a round hole in it, just big enough to admit the tube of the microscope, about two inches from the bottom, and equidistant from the two sides. Next cut off the two upper corners of the cardboard, and give them a pleasant-looking curve. Then cover the cardboard with black velvet, the commonest, which is not glossy, answers best, and your screen is made. Put the hole over the tube of the microscope, and let the screen rest on the little ledge or rim which forms an ornamental finish to most instruments. A piece of cork may be gummed at the back of the screen, so as to tilt it a little, and diminish its chance of coming into contact with that important organ the nose. This little contrivance adds to the clearness and brilliancy of objects, and is a great accommodation to the eyes.

One more oculistic memorandum, and we have done with this chapter. Do not stare at portions of objects that are out of focus, and consequently indistinct, as this injures the eyes more than anything. Remember the proverb, "None so deaf as those that won't hear," which naturally suggests for a companion, "None so blind as those that won't see." It is often impossible to get every object in the field in focus at one time;—look only at that which is in focus, and be blind to all the rest. This is a habit easily acquired, and is one for which our *natural* microscopes are exceedingly grateful; and every judicious observer desires to keep on the best terms with his eyes.

CHAPTER II.

JANUARY.

Visit to the ponds—Confervæ—Spirogyra quinina—Vorticella—Common Rotifer—Three divisions of Infusoria—Phytozoa—Protozoa—Rotifera—Tardigrada—Meaning of these terms—Euglenæ—Distinction between animals and vegetables—Description of Vorticellæ—Dark ground illumination—Modes of producing it—The Nucleus of the Vorticellæ—Methods of reproduction—Ciliated Protozoa—Wheel bearers or Rotifers—Their structure—The common Rotifer—The young Rotifer seen inside the old one—An internal nursery—"Differentiation" and "Specialisation"—Bisexuality of Rotifers—Their zoological position—Diversities in their appearance—Structure of their Gizzard—Description of Rotifers.

THE winter months are on the whole less favorable to the collection of microscopic objects from ponds and streams than the warmer portions of the year; but the difference is rather in abundance than in variety, and with a very moderate amount of trouble, representatives of the principal classes can always be obtained.

On a clear January morning, when the air was keen, but no ice had yet skinned over the surface of the water, a visit to some small ponds in an open field not far from Kentish Town provided entertainment for several days. The ponds were selected from their open airy situation, the general clearness of their water, and the abundance of vegetation with which they were adorned. Near the margin confervæ abounded, their tangled masses of hair-like filaments often matted together, almost with the closeness of a felted texture. At intervals, minute bubbles of air, with occasionally a few of greater size, indicated that the complex processes of vegetable life were actively going on, that the tiny plants were decomposing carbonic acid, dexterously combining the carbon—which we are most familiar with in the black opaque form of charcoal—to form the substance of their delicate translucent tissues, and sending forth the oxygen as their contribution to the purification of the adjacent water, and the renovation of our atmospheric air. This was a good sign, for healthy vegetation is favorable to many of the most interesting forms of infusorial life. Accordingly the end of a walking-stick was inserted among the green threads, and a skein of them drawn up, dank, dripping, and clinging together in a pasty-looking mass. To hold up a morsel of this mass, and tell some one not in the secrets of pond-lore that its dripping threads were objects of beauty, surpassing human productions, in brilliant colour and elegant form, would provoke laughter, and suggest the notion that you were poking fun at them, when you poked out your stick with the slimy treasure at its end. But let us put the green stuff into a bottle, with some water from its native haunt, cork it up tight, and carry it away for quiet examination under the microscope at home.

Here we are with the apparatus ready. We have transferred a few threads of the conferva from the bottle to the *live box*, spreading out the fine fibres with a needle, and adding a drop of water. The cover is then gently pressed down, and the whole placed on the stage of the microscope, to be examined with a power of about sixty. A light is thrown somewhat obliquely by the mirror through the object, the focus adjusted, and a beautiful sight rewards the pains. Our mass of conferva turns out to contain one of the most elegant species. Fine hair-like tubes of an organic material, as transparent as glass, are divided by partitions of the same substance into cylindrical cells, through which a slender ribbon of emerald green, spangled at intervals with small round expansions, is spirally wound. We shall call it the Spiral Conferva, its scientific name being *Spirogyra quinina.* Some other species, though less elegantly adorned, make a pleasing variety in the microscopic scene; and appended to some of the threads is a group of small crystal bells, which jerk up and down upon spirally twisted stalks. These are the "Bell Flower Animalcules" of old observers, the *Vorticellæ*, or Little Vortex-makers of the present day. Other small creatures flit about with lively motions, and among them we observe a number of green spindles that continually change their shape, while an odd-looking thing crawls about, after the manner of certain caterpillars, by bringing his head and tail together, shoving himself on a step, and then repeating the process, and making another move. He has a kind of snout, behind which are two little red eyes, and something like a pig-tail sticks out behind. This is the Common Wheel-bearer, *Rotifer vulgaris*, a favourite object with microscopists, old and young, and capable, as we shall see, of doing something more interesting than taking the crawl we have described.

A higher power, say one or two hundred, may be conveniently applied to bring out the details of the inhabitants of our live box more completely; but if the glasses are good, a linear magnification of sixty will show a great deal, with the advantage of a large field, and less trouble in following the moving objects of our search.

Having commenced our microscopic proceedings by obtaining some Euglenæ, Vorticellæ, and a Rotifer, we are in a position to consider the chief characteristics of three great divisions of infusoria, which will often engage our attention.

It is well known that animalcules and other small forms of being may be found in *infusions* of hay or other vegetable matter, and hence all such and similar objects were called *Infusoria* by early observers. Many groups have been separated from the general mass comprehended under this term, and it is now used in various senses. The authors of the 'Micrographic Dictionary' employ it to designate "a class of microscopic *animals* not furnished with either vessels or nerves, but exhibiting internal spherical cavities, motion effected by means of cilia, or variable processes formed of the substance of the body, true legs being absent." The objection to this definition is, that it to some extent represents theories which may not be true. That nerves are absent *all through the class* is an assumption founded merely upon the negative evidence of their not having been discovered, and the complete absence of "vessels" cannot be affirmed.

In the last edition of 'Pritchard's Infusoria,' to which some of our ablest naturalists have contributed, after separating two groups, the Desmids, and the Diatoms, as belonging to the vegetable world, the remainder of the original family of infusoria are classified as *Phytozoa*, *Protozoa*, *Rotifera*, and *Tardigrada*. We shall explain these hard names immediately, first remarking that the Desmids and the Diatoms, concerning whom we do not intend to speak in these pages, are the names of two groups, one distinctly vegetable, while the other, although now generally considered so, were formerly held by many authorities to be in reality animal. The Desmids occur very commonly in fresh water. We have some among our Confervæ. They are most brilliant green, and often take forms of a more angular and crystalline character than are exhibited by higher plants. The Diatoms are still more common, and we see before us in our water-drop some of their simplest representatives in the form of minute boats made of silica (flint) and moved by means still in dispute.

Leaving out the Desmids and Diatoms, we have said that in Pritchard's arrangement the views of those writers are adopted who divide the rest of the infusoria into four groups, distinguished with foreign long-tailed names, which we will translate and expound. First come the *Phytozoa*, under which we recognise our old acquaintance *zoophyte* turned upside down. *Zoophytes* mean animal-plants, *Phytozoa* mean plant-animals. We shall have by-and-bye to speak of some of the members of this artificial and unsatisfactory group, and postpone to that time a learned disquisition on the difference between animals and plants, a difference observable enough if we compare a hippopotamus with a cabbage, but which "grows small by degrees, and beautifully less," as we contemplate lower forms.

After the *Phytozoa* come the *Protozoa*, or first forms in which animality is distinctly recognised. Under this term are assembled creatures of very various organization, from the extreme simplicity of the *Proteus* or *Amœba*, a little lump of jelly, that moves by thrusting out portions of its body, so as to make a sort of extempore legs, and in which no organs can be discerned,[2] up to others that are highly developed, like our *Vorticellæ*. This group is evidently provisional, and jumbles together objects that may be widely separated when their true structure and real affinities are discerned.

[2] In some kinds and in some stages of growth this is not strictly true.

Following the *Protozoa*, come the *Rotifera*, or Wheel-bearers, of which we have obtained an example from our pond, and whose characteristics we shall endeavour to delineate when our specimen is under view; and last in the list we have the *Tardigrada*, "Slow-steppers," or Water Bears, queer little creatures, something like new-born puppies, with a double allowance of imperfect feet. These, though somewhat connected with the rotifers, are considered to belong to a low division of the arachnida (spiders, &c.).

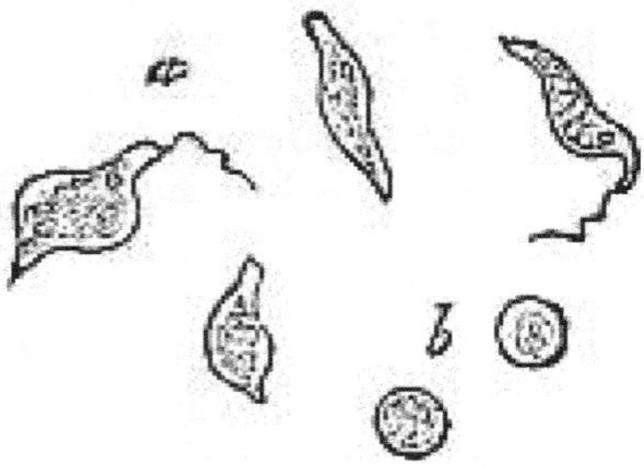

a, motile; and b, resting condition of Euglenæ.

Feeling that we must be merciful with the long-tailed words and explanations of classification, we reserve further matter of this kind for the opportunities that must arise, and direct our attention to living forms by watching the *Euglenæ* which our water-drop contains. We have before us a number of elegant spindle-shaped bodies, somewhat thicker in front than behind, and in what may be called the head there glitters a brilliant red speck, commonly called an *eye-spot*, although, like the eyes of potatoes, it cannot see. Round this eye-spot the tissues are clear, like glass; but the body of the creature is of a rich vegetable green, which shines and glistens as it catches the light. Some swim rapidly with a rollicking motion, while others twist themselves into all manner of shapes. Now the once delicate spindle is oddly contorted, now it swells out in the middle, like a top, and now it rolls itself into a ball. The drawings will afford some idea of these protean changes, but they must be seen before their harlequin character can be thoroughly appreciated. Some of the specimens exhibit delicate lines running lengthwise, and taking a spiral twist as the creature moves about; but in none can any mouth be discerned, and their antics, although energetic and comical, afford no certain indications of either purpose or will. What are they? animals or vegetables? or something betwixt and between?

The first impression of any casual observer would be to declare in favour of their animality; but before this can be settled, comes the question, what is an animal, and how does it differ from a vegetable? and upon this the learned do by no means agree. One writer considers the presence of *starch* in any object a proof that it belongs to the dominions of Flora, while another would decide the issue by ascertaining whether it evolves oxygen and absorbs carbon, as most plants do, or whether it evolves carbon and absorbs oxygen, as *decided* animals do. Dr. Carpenter asserts that the distinction between *Protophyta* and *Protozoa* (first or simplest plants and animals), "lies in the nature of their food, and the method of its introduction, for whilst the *Protophyte* obtains the materials of its nutrition from the air and moisture that surround it, and possesses the power of detaching oxygen, hydrogen, carbon, and nitrogen from their previous binary combinations, and of uniting them into ternary and quaternary organic compounds (chlorophyll, starch, albumen, &c.), the simplest *Protozoa*, in common with the highest members of the animal kingdom, seems utterly destitute of any such power, makes, so to speak, a stomach for itself in the substance of its body, into which it injects the solid particles that constitute its food, and within which it subjects them to a regular process of digestion."

Unfortunately it is very difficult to apply this simple theory to the dubious objects which lie on the border-land of the animal world, and no other theory that has been propounded appears to meet all cases. Some naturalists do not expect to find a broad line of demarkation between the two great divisions of living things, but others characterise such an idea as "unphilosophical," in spite of which, however, we incline towards it.

Mr. Gosse, whose opinion is entitled to great respect, calls the *Euglenæ* "animals" in his 'Evenings with the Microscope;' but from the aggregate of recorded observations it seems that they evolve oxygen, are coloured with the colouring matter of plants, reproduce their species in a manner analogous to plants, and have in some cases been clearly traced to the vegetable world. It is, however, possible that some *Euglenæ* forms may be animal and others vegetable, and while their place at nature's table is being decided, they must be content to be called *Phytozoa*, which, as we have before explained, is merely *Zoophyte* turned upside down.

Some authorities have thought their animality proved by the high degree of contractility which their tissues evince. This, however, cannot go for much, as all physiologists admit contractility to belong to the vegetable tissues of the sensitive plant, "Venus' Fly-trap," &c., and a little more or less cannot mark the boundary between two orders of being.

We shall have occasion again to notice the *Protophytes*, and now pass to the *Protozoa*, of which we have a good illustration in the *Vorticella* already spoken of. In the group before us a number of elegant bells or vases stand at the end of long stalks, as shown at the top of the frontispiece, while round the tops of the bells, the vibrations of a wreath or cilia produce little vortices or whirlpools, and hence comes the family name. This current brings particles of all sorts to the mouth near the rim of the bells, and the creature seems not entirely destitute of power to choose or reject the morsels according to its taste. Every now and then the stalk of some specimen is suddenly twisted into a spiral, and contracted, so as to bring the bell almost to the ground. Then the stem gracefully elongates again, and the cilia repeat their lively game.

The general effect can be seen very well by a power of about sixty linear, but one of them from one to two hundred is necessary to bring out the details, and a practised observer will use still more magnification with good effect. They should be examined by a moderately oblique light, or most of the cilia are apt to be rendered invisible, and also by *dark ground* illumination. This may be accomplished in a well-made microscope by turning the mirror quite out of the plane of the axis of the instrument, that is to say, on one side of the space the body would occupy if it were prolonged. By this means, and by placing the lamp at an angle with the mirror, that must be learnt by experiment, all the light that reaches the eye has first passed through the object, and is refracted by it out of the line it was taking, which would have carried it entirely away. Or the object may be illuminated by an apparatus called a *spotted lens*, which is a small bull's-eye placed under the stage, and having all the centre of its face covered with a plaster of black silk. In this method the central or direct rays from the mirror are obstructed, but those which strike the edge of the bull's-eye are bent towards the object, which they penetrate and illuminate if it is sufficiently transparent and refractive. Another mode of dark ground illumination is by employing an elegant instrument called a *parabolic illuminator*, which need not be described.

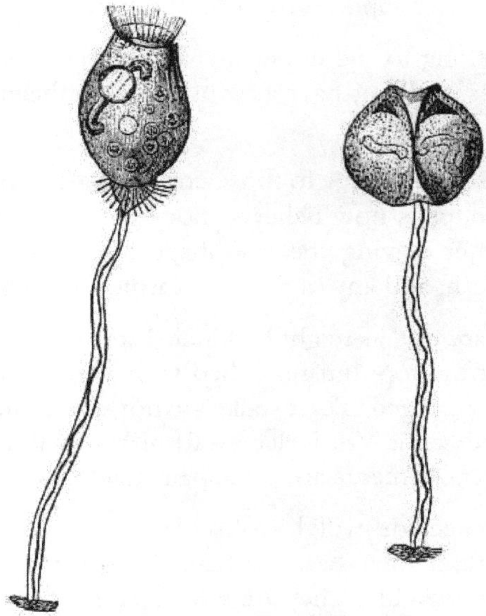

Left: Vorticella, with posterior circlet of cilia in process of separation—Stein.
Right: Vorticella in process of self-division. A new frontal wreath in formation in each of the semi-lunar spaces.

Different specimens and species of *Vorticellæ* vary in the length of their bells from one three or four thousandth to one hundred and twentieth of an inch, and when they are tolerably large, the dark ground illumination produces a beautiful effect. The bells shine with a pearly iridescent lustre, and their cilia flash with brilliant prismatic colours.

Left: Vorticella microstoma, showing alimentary tube, ciliated mouth, and formation of a gemma at the base, 300 linear.—Stein.

Right: Vorticella microstoma, the encysted animal protruding through a supposed rupture of the tunic.

The *Vorticellina* belong to the upper division of the *Protozoa*—the *ciliata*, or ciliated animalcules, and they have a mouth, an œsophagus, and an orifice for the exit of their food.

Many observers used to ascribe to those creatures a complete intestinal canal, but such an apparatus is now believed not to exist in any of the Infusoria. Food particles, after leaving the œsophagus, are thrust forward into the sarcode, or soft flesh, and any cavity thus formed acts as a stomach.

The bells or cups are not, as might be fancied from a casual inspection, open like wineglasses at the top, but furnished with a retractile disk or cover, on which the cilia are arranged. Their stalks are not simple stems, but are hollow tubes, which in the genus Vorticella are furnished with a muscular band, by whose agency the movements are principally made.

Some of the Vorticellids will be observed to leave their stalks, having developed cilia round their base, and may be seen to swim about in the enjoyment of individual life. They are also capable of becoming *encysted*, that is, of secreting a gelatinous cover.

Encysted Vorticella, showing the obliteration of special organs by the advancement of the process.—Pritchard.

These changes are exhibited in the annexed cuts, which are copied from known authorities. By careful observation of the bodies of Vorticellids, a contractile vesicle may be observed, which appears to cause a movement of fluids, that is probably connected either with respiration or secretion.

Another piece of apparatus in this family, but not confined to it, is the so-called *nucleus*, which in this case is of a horseshoe shape and granular texture, and greater solidity than the surrounding parts. The functions of this organ formed the subject of various conjectures, but it is now generally held to be an ovary.

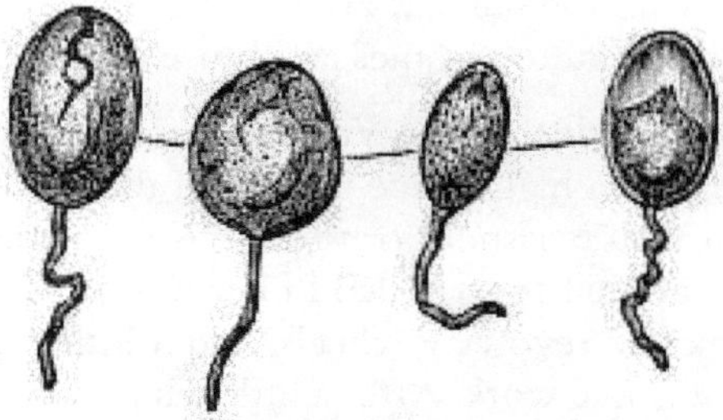

Vorticella microstoma, in process of encystment, 300 linear; in the last the inclosing tunic is plainly developed.—Stein.

In common with many of the lower animals, the Vorticellids have three ways of multiplying their race. One by *fission*, or division of their bodies: another by *buds*, somewhat analogous to those of plants; and another by reproductive germs. These processes will come again under our notice, and we shall leave the Vorticellids for the present by observing that if they are fed with a very small quantity of indigo or carmine, the vacuoles or spaces, into which their nutriment passes, will be clearly observed. Ehrenberg thought in these and similar creatures that every vacuole was a distinct stomach, and that all the stomachs were connected by an intestinal canal; hence his name *Polygastrica*, or many stomached. In these views he has not been followed by later observers, and it is probable he was misled, partly by pushing the process of reasoning from the analogies of higher animals much too far, and partly by the imperfection of the glasses he employed.

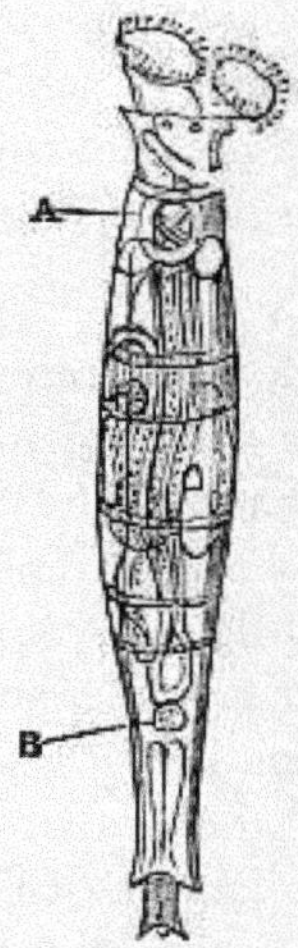

Rotifer vulgaris.—A, mouth, or gizzard; B, contractile vesicle.—Micrographic Dictionary.

N.B.—When the cilia and tail part are retracted, and the body shortened, the creature assumes an obtuse oval form.

Having thus briefly considered the Vorticellids we must turn to the wheel-bearer, who belongs to a higher race than even the ciliated *Protozoa.* We left her crawling about with her snout or proboscis protruded, but now she has moored herself by her tail-foot, pulled in her nose, and put out two groups of cilia, which look like revolving wheels, and a little below them is seen a gizzard in a state of active work. After a little while she swims away with her wheels going, and her tail, forked at the end, is found to be telescopic, or capable of being pulled in and out. As the cilia play, the neighbouring water is agitated, and the multitudes of small objects are brought by the whirlpools within her ravenous maw. But the strangest thing of all is that inside her body is seen a young one; in this case a large and fine infant, which, like "a chip of the old block," imitates the parental motions, thrusts forth its cilia and works its gizzard.[3] In other genera the eggs are hatched externally, but this one is ovoviparous, and carries its nursery inside.

[3] This was met with in the summer, but is described here to avoid repetition. I do not know whether the eggs are hatched in very cold weather.

A very slight investigation is sufficient to show that in the wheel-bearer we have made a great advance towards a higher organization than we discovered in the preceding creatures. We witness what the learned call a "differentiation" of parts and tissues, and a "specialization" of organs. The head is plainly distinguishable from the body, the skin or integument is distinctly different from the internal tissues, behind the eyes we can detect a nervous ganglion or miniature brain, the gizzard is a complicated piece of vital mechanism, such as we have not met with before, and in various parts of the transparent inside we see organs to which particular functions are assigned.

It was at one time thought that Rotifers were hermaphrodite—uniting both sexes in one body—but that idea is now generally abandoned, for in many species the males have been discovered, and the fair sex may be gratified to hear that they are without doubt the "inferior animals." Their function is simply to assist the female in producing young, and as this can be quickly accomplished, their lives are short, and they are not supplied with the gizzard and digestive apparatus, which their lady-loves possess. Much discussion has taken place as to the rank which the Rotifers hold in the animal kingdom, some naturalists thinking them relations of the crabs, and others believing them to belong to the family of the worms. Professor Huxley, who adopts the latter view, which has the most friends, groups the lower *Annulosa* together under the name of *Annuloida*, in which he includes *Annelides*, or worms of various kinds, the *Echinodermata* (or "spine skins," among which are the star-fish and sea hedgehogs), and some other families. He considers the Rotifers to be "the permanent forms of Echinoderm larvæ." This does not mean that they were ever produced by Echinoderms, and had their development checked, but that they resemble them in organization, and illustrate a general law, observable in animated beings, namely, that the lower creatures are like the imperfect stages of higher animals, and that all things are formed according to general principles, and exhibit a uniformity of plan.

Mr. Gosse adopts a different view, and while admitting a connection between the Rotifers and the worms, adduces important reasons for associating them with the insects.

Leaving zoologists to settle their position, we may remark that the Rotifers form a very numerous family, presenting very great diversities of structure, some of the most interesting of which we shall meet with in the course of our rambles; but they all possess a gizzard, which, though differing in complexity, is throughout formed upon the same principle, and that we must now explain.

We have called the masticatory apparatus of the Rotifers a *gizzard*; but Mr. Gosse, who has done most to elucidate its structure, contends that it is a *mouth*; and in some species it is frequently protruded, and used like the mouth of higher animals. Taking one of the most typical forms of this organ, and drawing our illustrations from Mr. Gosse's admirable paper in the "Transactions of the Royal Society," we may describe it, when completely developed, as consisting of three lobes, having a more or less rounded form. The eminent naturalist we have named calls the whole organ the *mastax*, and states that it is composed of dense muscular fibre. The tube which leads down to it he designates the "buccal (mouth) funnel," and the tube that issues from it, and conveys the food to the digestive sac or stomach, he calls the *œsophagus*, in conformity with the nomenclature applied to creatures whose mouths are in the usual place. Inside the mouth-gizzard are placed two organs, which work like hammers, and which Mr. Gosse therefore names *mallei*. The hammers work against a sort of anvil, which is called *incus*, the Latin for that implement. Each hammer consists of two portions articulated by a hinge joint. The lower portion, the *manubrium*, or handle, gives motion to the upper portion, which from its shape is named the *uncus*, or hook. The *unci* are furnished with finger-like processes of teeth, which vary in number. There are five or six in the best developed specimens. These hooks or teeth work against each other, and against the *incus*, or anvil, which consists of distinct articulated portions, of which the principal are two *rami*, or branches, jointed so that they can open and close like a pair of shears. These two rest upon the third portion, which is called the *fulcrum*. Some faint idea of the working of the toothed hammers may be obtained by rubbing the knuckles of both hands together, but the motion is more complicated, and the *rami* play their part in the trituration of the food. Mr. Gosse states that when an objectionable morsel has got as far as this mouth-gizzard, "it is thrown back by a peculiar scoop-like action of the *unci*, very curious to witness." The foregoing diagram will help the reader to comprehend this description, but no opportunity should be lost for viewing this remarkable organ busy at work in the living animals.

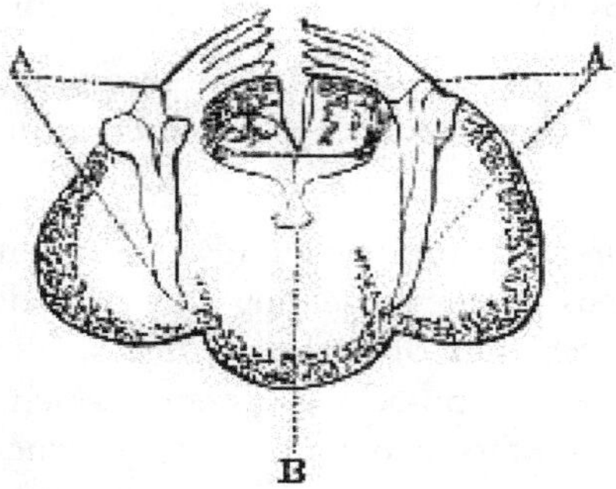

Gizzard of Notomata.

The respiration of the Rotifers is supposed to be effected by the passage of water through vessels running round them, and called the "water vascular system," and in addition to their eyes, which often disappear in adult specimens, the organ we described as standing out like a pig-tail, as our acquaintance crawled along, is thought to act as an *antenna*, or feeler, and brings its possessor in further relation to the external world. It is also called the *calcar*, or spur, and is furnished with cilia or bristles at its extremity.

Sometimes the particles swallowed by the Common Rotifer are large enough for their course to be traced, but there is frequently a great commotion and grinding of the gizzard, without any appreciable cause, although doubtless something is taken in, and when the creature is tired, or has had enough, we see both head and tail retracted, and the body assumes a globular form. In another chapter, when viewing a Philodine, we shall see how in the family to which the Common Rotifer belongs, the gizzard departs from the perfect type.

CHAPTER III.

FEBRUARY.

Visit to Hampstead—Small ponds—Water-fleas—Water-beetle—Snails—Polyps—Hydra viridis—The dipping-tube—A glass cell—The Hydra and its prey—Chydorus sphæricus and Canthocamptus, or friends and their escapes—Cothurnia—Polyp buds—Catching Polyps—Mode of viewing them—Structure of Polyps—Sarcode—Polyps stimulated by light—Are they conscious?—Tentacles and poison threads—Paramecium—Trachelius—Motions of Animalcules, whether automatic or directed by a will—Their restless character.

IT has been a bitterly cold night, and as the sun shines on a clear keen morning, and glistens in the hoar-frost which covers the trees, it might seem an unpropitious time for visiting the ponds, in search of microscopic prey. We will, however, try our luck, and take a brisk trot to the top of Hampstead Heath, where the air is still keener, and the ice more thick. Arriving at the highest point, London appears on one side enveloped in its usual great coat of smoke, through which St. Paul's big dome, with a score or two of towers and steeples, can be dimly made out; while looking towards Harrow-on-the-Hill, or Barnet, we see the advantage of country air in the sharpness with which distant objects cut the blue sky. We leave the large ponds for another time, and hunt out the little hollows among the furze and fern. One looks promising from the bright green vegetation to be discovered under the sheet of ice, which is almost firm enough to bear human weight.

Breaking a convenient hole we hook up some of the water-plants, and place them in a wide-mouthed vial, which we fill with water, and cursorily examine with a pocket-lens. Some water-fleas briskly skipping about, and a beautiful little beetle, with an elegant dotted pattern on his brown back, and a glistening film of air covering his belly, show that we have not been unsuccessful, although we must wait till we get home to know the extent of our findings, among which, however, we can only discern the graceful spiral shell of a small water-snail, the *Planorbis*.

Arriving at home the bottle was left undisturbed for some hours in a warm light place, and then on being examined several specimens of that beautiful polyp, the *Hydra viridis*, were seen attached to the glass, and spreading their delicate tentacles in search of prey. One of the polyps is carefully removed by the *dipping-tube*, a small glass tube, open at both ends. The forefinger is placed upon the top, and when the other end is brought over the object the finger is raised for an instant, and as the water rushes in the little hydra comes too, and is placed in a glass cell, about half an inch wide, and one tenth of an inch deep. These cells are obtained from the opticians, and cemented with varnish or marine glue to an ordinary glass slide. After an object has been placed in one of them, a little water is taken up in the dipping-tube, and the cell filled until the fluid stands in a convex heap above its brim. We then select around glass cover, and press it gently on the walls of our cell. A few drops of superfluous water escape, and we have the cell quite full, and the cover held tight by force of the capillary attraction between the water and the glass.

Hydra viridis with developed young one, and bud beginning to sprout.

The polyp deposited in one of these water cages is then transferred to the stage of the microscope, and its proceedings watched. At first it looks like a shapeless mass of apple-green jelly. Soon, however, the tail end of the creature is fixed to the glass, the body elongates, and the tentacles (in this case eight) expand something after the manner of the leaves of a graceful palm.

By accident two small Water Fleas were imprisoned with the polyp, and one (a shrimp-like looking creature, carrying behind her a great bag of eggs) came into contact with the tentacles, and seemed paralysed for a time. The hydra made no attempt to convey the captive to its mouth, but held it tight until another Water Flea, a round merry little fellow (*Chydorus sphæricus*), came to the rescue, and assisted *Canthocamptus* to escape by tugging at her tail. This friendly action may not have been prompted by the intelligence which seemed to suggest it, but those who have kept tame soldier-crabs and prawns in an aquarium, will not be indisposed to attribute to the crustaceans more brains than they have usually credit for. It must, however, be confessed that the subsequent conduct of Mrs. Canthocamptus did not indicate the possession of much prudence, for she learnt no lesson from experience, but repeatedly swam against her enemy's tentacles, suffered many captures, and only escaped being devoured through the indifference, or want of appetite, which the polyp evinced.

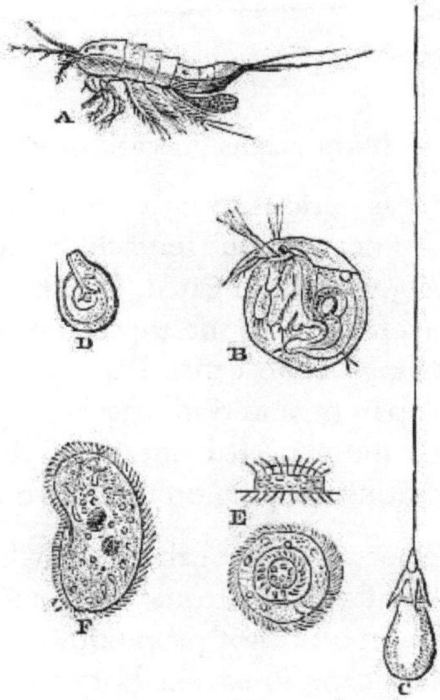

A, Canthocamptus minutus; B, Chydorus sphæricus; C and D, Capsules and poison-thread of polyp; E, Tricodina pediculus, side view and under view; F, Kerona polyporum.—Microg. Dict.

On the body of the *Canthocamptus* were some small transparent vases or bottles, containing living objects, which sprang up and down. These were members of the *Vorticella* family, called *Cothurnia*, and will be hereafter described.

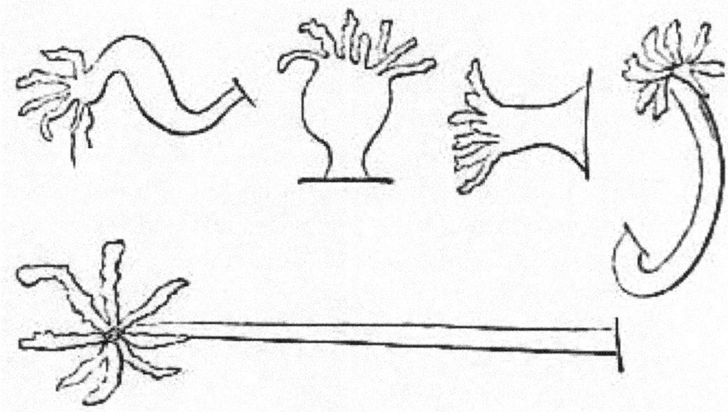

Hydra viridis, in various shapes.

Watching the hydra it was curious to note the changes of form which these creatures are able to assume. Now the tentacles were short and thick, and the body squat; now the body was elongated, like the stem of a palm tree, and the tentacles hung gracefully from the top. From some of the polyps little round buds were growing, while other buds were already developed into miniature copies of the parent, and only attached by a slender stalk. In a few days many of these left the maternal side, fixed their own little tails to the glass, and commenced housekeeping on their own account.

Polyps may be obtained at all times of the year by bringing home duckweed, conferva, and other water-plants from the ponds. Some hauls may be unsuccessful, but if one pond is not propitious others should be tried. The plants should be put in a capacious vessel of water, and placed in the light, where, if polyps be present, they will show themselves within twenty-four hours, either attached to the sides of the glass, or hanging from the plants, or suspended head downwards from the upper film of the water. They are elegant objects, and may be kept without difficulty for some weeks. After being confined in a small quantity of water for purposes of examination, they should be carefully replaced in the larger vessel, and may thus be used again and again without suffering any injury. A low power—a three or two-inch glass—or a one-inch, reduced by employing the erector—is the most convenient for examining the whole creature, but higher powers are necessary to make out its minute structure. They should be viewed with direct and oblique light, as transparent and also as opaque objects. In the latter case the "Lieberkuhn," or polished silver speculum, is convenient, and if the microscope is not furnished with Lister's dark wells, a small piece of black paper may be stuck behind the object, by simply wetting it with the tongue.[4]

[4] The side silver reflector is useful for illuminating such objects.

Although the polyps are remarkable for the simplicity of their organization, they do not the less exhibit the wonderful nature of animal life. Their bodies are composed of the substance, called *sarcode*, in which is imbedded a colouring matter resembling that in the leaves of plants; every part possesses irritability and contractility, and they are very sensitive to the stimulus of light. The outer layer of their bodies is harder than the inner layer. These layers are severally called *ectoderm* and *endoderm*. They may be cut and grafted like trees, and if turned inside out, the new inside digests and assimilates as well as the old. Whether any form of consciousness can belong to creatures which have no distinct nervous system is open to doubt, but it would seem probable from their movements that food and light afford them something like a pleasurable sensation in a very humble degree. If we were sufficiently acquainted with the secrets of molecular combination we might discover that the various functions of these simple organisms were discharged by different *particles*, although it is only in higher creatures that muscular particles are aggregated into muscles, or nerve particles into nerves.

Having examined the general appearance and proceedings of the hydra, let us cut off a tentacle, or take a small specimen and gently crush it by pressing down the cover of the live box, and place the object so prepared under a power of about three hundred linear. If we then illuminate it with a moderate quantity of oblique light, we shall discover round the edge of the tentacle a number of small cells or capsules, from some of which a very slender wire or thread will be emitted.[5] These are the stinging organs of the polyp, and resemble those which Mr. Gosse has so ably elucidated in the sea anemones. Some writers have endeavoured to show that they are not stinging organs at all, but so large an amount of evidence to the contrary is accumulated in Mr. Gosse's 'Actinologia Britannica,' that no reasonable doubt remains. The stinging capsules of the polyp are shown in the annexed sketch, and also the way in which they are employed, for it fortunately happened that on exposing one of the hydras to pressure in the live box, a small worm (*Anguillula*) escaped, which had been pierced with the minute weapons which are supposed to convey a poison into the wound. The authors of the 'Micrographic Dictionary' think that the prongs of the forks, which will be seen to point upwards in the sketch,[6] are springs, and occupy a reversed position in the capsule cells, and that their function is to throw out the threads. However this may be, the polyps, and similarly endowed creatures, have the power of darting out their poison threads with considerable force, and Mr. Gosse found that the anemone was able to pierce a thick piece of human skin.

[5] See illustration above.

[6] See illustration below.

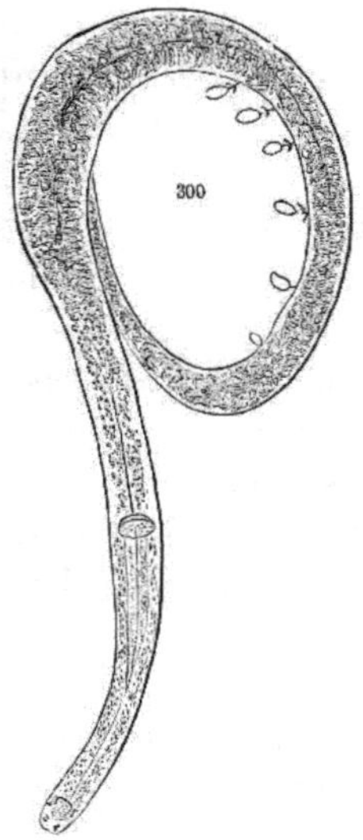

Anguillula pierced by stinging organs of the Hydra viridis.

The same excellent observer attributes the emission of the anemone poison threads, which he considers hollow, to the injection of a fluid. In their quiescent state, he thinks they are drawn in, like the finger of a glove, and are forced out as the liquid enters their slender tubes. Possibly the polyp stinging organs may have the same structure.

Notwithstanding their dangerous weapons, polyps are often infested with a parasite, the *Trichodina pediculus*, as shown in Fig. E, page 49, and it must happen that either this visitation is not disagreeable, or that the Trichodina is not influenced by the poison.

As the plants in the bottles decayed, some of the animalcules died off and others appeared. In one bottle, containing decaying chara, *Paramecia* abounded. The *Paramecia*, of which there are various species, have always been favourite objects with microscopists. The Germans call them "slipper animalcules," and they vary in size from 1—96" [7] to 1—1150". They are flat rounded-oblong creatures, with a distinct integument or skin, "through which numerous vibratile cilia pass in regular rows."[8] They are furnished with a distinct mouth, and adult specimens exhibit star-shaped contractile vesicles in great perfection.

[7] The usual mode of giving dimensions is by fractions thus expressed: 1—96" means one ninety-sixth of an inch.

[8] 'Micrographic Dictionary.'

The swarm of specimens before us belong to one species, *Paramecium aurelia*, the *Chrysalis animalcule*, and they crowd every portion of the little water-drop we have taken up, and examined with a power of about one hundred linear. When they are sufficiently quiet a power of about four hundred may be used with advantage, and Pritchard recommends adding a little indigo and carmine to the water, in order to see the cilia more clearly, or rather to render their action more plain. The cilia are disposed lengthwise, and Ehrenberg counted in some rows sixty or seventy of them, making an aggregate of three thousand six hundred and forty organs of motion in one small animated speck. This number seems large, but although we have never performed the feat of counting them, we should have expected it to prove much greater. Unlike most animalcules they are susceptible of being preserved by drying upon glass, and we subjoin a figure from Pritchard, of one thus treated, in which the star-shaped vesicles are clearly seen. These curious organs communicate with other vessels, and, as we have previously stated, are probably connected with respiration and excretion.

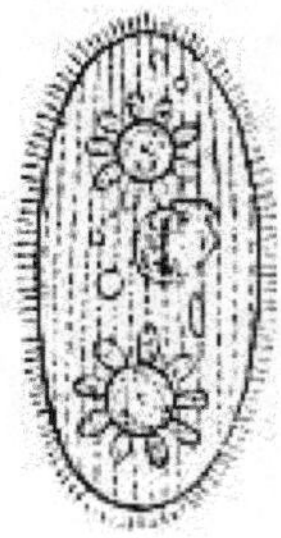

Paramecium aurelia. A dried specimen showing the vesicles.—Pritchard.

The genus *Paramecium* is now confined to those creatures which exhibit rows of longitudinal cilia of uniform length, which are destitute of hooks, styles, or other organs of motion than the cilia, which have a lateral mouth, and no eye-spots. One mode of increase is by division, which may be easily observed; another is through the formation of true eggs as traced by Balbiani.

Another of the treasures from the pond was a species of *Trachelius*, or long-necked ciliated animalcule, which kept darting in and out of a slimy den, attached to the leaf of a water-plant. The body was stout and fish-shaped, the tail blunt, and the neck furnished with long conspicuous cilia, which enabled the advancing and retreating movements to be made with great rapidity. The motions of this creature exhibit more appearance of purpose and design than is common with animalcules, but in proportion as these observations are prolonged, the student will be impressed with the difficulty of assuming that anything like a reasoning faculty and volition, is proved by movements that bear some resemblance to those of higher animals, whose cerebral capacities are beyond a doubt. It is, however, almost impossible to witness motions which are neither constant nor periodic, without fancying them to be dictated by some sort of intelligence. We must, nevertheless, be cautious, lest we allow ourselves to be deceived by reasoning so seductive, as the vital operations of the lowest organisms may be merely illustrations of blind obedience to stimuli, in which category we must reckon food, and until we arrive at forms of being which clearly possess a ganglionic system, we have no certainty that a real will exists, even of the simplest kind; and perhaps we must go still higher before we ought to believe in its presence.

Ehrenberg was much struck with the restless character of many infusoria—whether he looked at them by day or by night, they were never still. In fact their motions are like the involuntary actions which take place in the human frame; and if attached to their bodies we observe cilia that never sleep, the living membrane of some of our own organs, the nose, for example, is similarly ciliated, and keeps up a perpetual though unconscious work.

CHAPTER IV.

MARCH.

Paramecia—Effects of Sunlight—Pterodina patina—Curious tail—Use of a Compressorium—Internal structure of Pterodina—Metopidia—Trichodina pediculus—Cothurnia—Salpina—Its three-sided box—Protrusion of its gizzard mouth.

THE *Paramecia*, noticed in the last chapter, have increased and multiplied their kind without any fear lest the due adjustment between population and food should fail to be preserved. A small drop of the scum from the surface of the water in their bottle is an astounding sight. They move hither and thither in countless numbers, seldom jostling, although thick as herrings in a tub, and in many portions of the field the process of self-fissure, or multiplication by division, is going on without any symptoms of discomfort on the part of the parent creature. This is an interesting sight, but we will not linger over it, for the sun is shining, and there is enough warmth in the air to make it probable that the ponds will be more prolific than in the cold winter months. Sunshine is a great thing for the microscopic hunter; it brings swarms of creatures to the surface, and the Rotifers are especially fond of its genial beams. Even if we imitate it by a bright lamp, we shall attract crowds of live dancing specks to the illuminated side of a bottle, and may thus easily effect their capture by the dipping-tube.

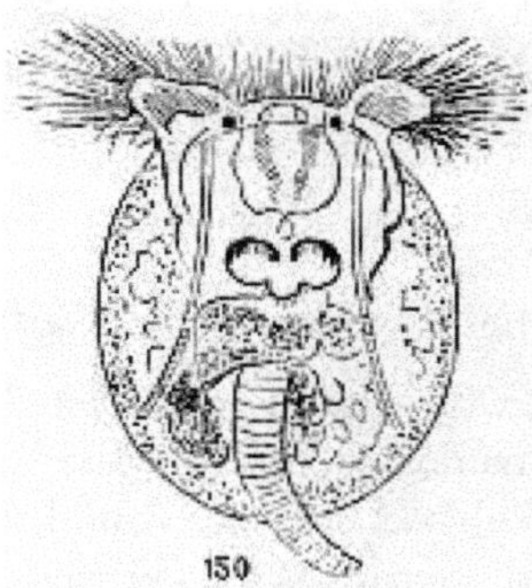

150

Pterodina patina.

This year the March sunshine was not lost, for on the third of that month I obtained a bottleful of conferva from a pond about a mile from my house, and lying at the foot of the Highgate hills. Water-fleas were immediately discovered in abundance, together with some minute worms, and a ferocious-looking larva covered with scales; but what attracted most attention was a Rotifer, like a transparent animated soup-plate, from near the middle of which depended a tail, which swayed from side to side, as the creature swam along. The head exhibited two little red eyes; two tufts of cilia rowed the living disk through the water, and the gizzard worked with a rapid snapping motion, that left no doubt the ciliary whirlpools had brought home no slender stores of invisible food. Sometimes the end of the tail acted as a sucker, and fixed the animal tightly to the glass, when the wheels were protruded, and the body swayed to and fro. Then the sucker action ceased, and as the creature swam away, a tuft of cilia was thrust out from the extremity of the tail. A power of one hundred linear was sufficient to enable the general nature of this beautiful object to be observed, but to bring out the details, much greater amplification was required, and this would be useless if the little fidget could not be kept still.

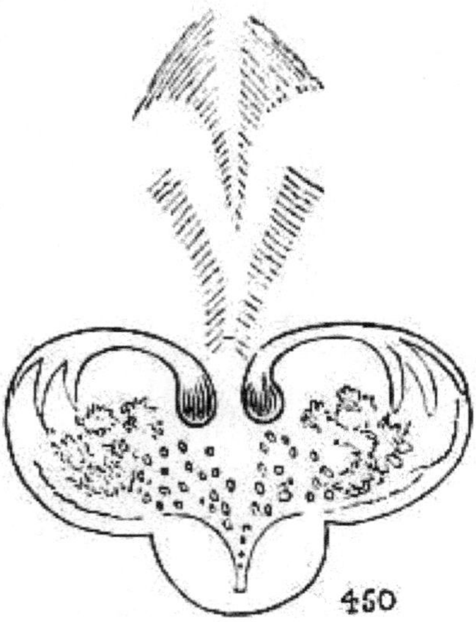

450 Pterodina patina—gizzard.

The size of the creature, whose name we may as well mention was *Pterodina patina*, rendered this practicable, but required some care. The longest diameter of the body, which was not quite round, was about 1—120", so that it was visible to the naked eye, and as a good many were swimming together, one could be captured without much difficulty, and transferred with a very small drop of water to the live-box. Then the cover had to be put on so as to squeeze the animal just enough to keep it still without doing it any damage, or completely stopping its motions. This was a troublesome task, and often a little overpressure prevented its success.

Some observers always use in these cases an instrument called a *compressorium*, by which the amount of pressure is regulated by a lever or a fine screw; but whether the student possess one or not, he should learn to accomplish the same result by dexterously manipulating a well-made live-box. We will suppose the *Pterodina* successfully caged, and a power of about one hundred and fifty linear brought to bear upon her, for our specimen is of the "female persuasion." This will suffice to demonstrate the disposition and relation of the several parts, after which one of from four hundred to five hundred linear may be used with great advantage, though in this case the illumination must be carefully adjusted, and its intensity and obliquity frequently changed, until the best effect is obtained.

We find, on thus viewing the Pterodina, that it is a complex, highly organized creature, having its body protected by a *carapace*, like the shell of a tortoise, but as flexible as a sheet of white gelatine paper, which it resembles in appearance. Round the margin of this carapace are a number of little bosses or dots, which vary in different individuals. The cilia are not disposed, as at first appeared, in two separate and distinct disks, but are continuous, as in the annexed sketch. Down each side are two long muscular bands, distinctly *striated*, and when they contract, the ciliary apparatus is drawn in. As this contraction takes place, two apparently elastic bands, to which the ciliary lobes are attached, are bent downwards, till they look like the C springs behind a gentleman's carriage; and they regain their former position of slight curvature, when the cilia are again thrust out.

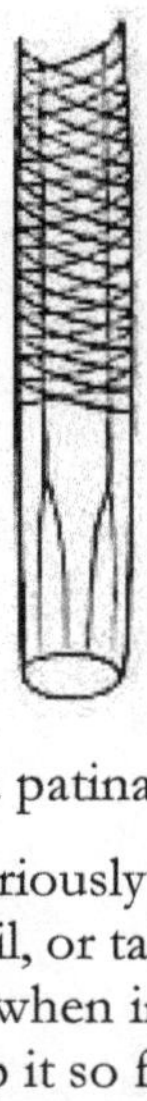

Pterodina patina—tail-foot.

The gizzard is three-lobed, and curiously grasped by forked expansions of the handles of the hammers. The tail, or tail-foot, can be withdrawn or thrust out at the will of the creature; and when in a good position for observation, a slight additional pressure will keep it so for examination. Delicate muscular longitudinal bands, forked towards the end of their course, supply the means of performing some of its motions, and one, or perhaps two, spiral threads extend through the upper half of its length, and either act as muscles, or as elastic springs for its extension. The intestines and other viscera are clearly exhibited, and a strong ciliary action conducts the food to the gizzard-mouth.

To return to the tail. One spiral fibre is easily discovered; but I have often, and at an interval of months, seen the appearance of two, and am in some doubt whether this was a deception, arising from the compression employed, or was a genuine indication.

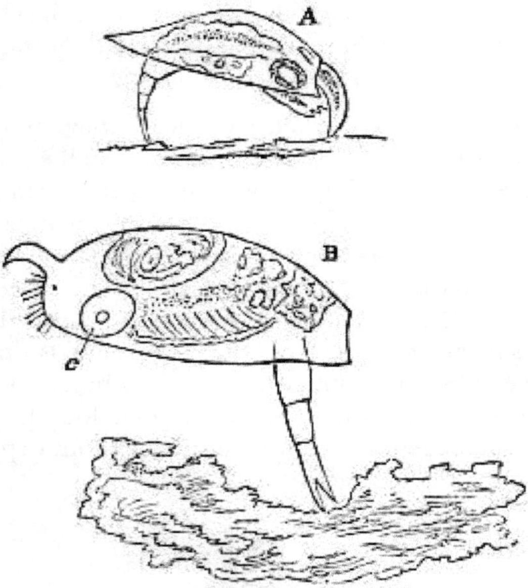

A. Metopidia acuminata, as drawn by Mr. Gosse. B. Specimen as seen and described in text. c. Mouth or gizzard.

Where this Rotifer occurs I have usually found it plentiful, but unfortunately could obtain no constant supplies after I had determined to make a special study of the remarkable tail, which is much more complicated than I have described. The *Pterodina* lived for some time in captivity, and for a week or two I could obtain them from my glass tank. They were likewise to be found for some weeks in the same part of the pond, but not all over it, until one day not a single specimen could be discovered, notwithstanding a persevering search nor was I afterwards able to get any from that pond during the remainder of the year.

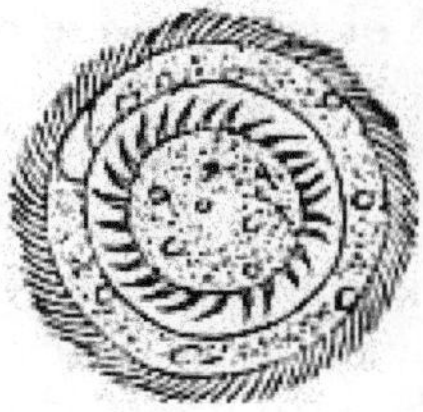

260

Trichodina pediculus.

Several other Rotifers, with and without carapaces, were among the same mass of confervæ, among them a *Metopidia*, with a firm shell, a forked jointed tail, and a projection in front which worked like a pickaxe among the decaying weed. There were likewise specimens of the long-necked animalcules (Trachelii), groups of Vorticella, some specimens of Volvox, and a small *Trichodina pediculus*, which, when magnified two hundred and sixty linear, was about the size of a sixpence and equally round. The edge was beautifully fringed with a circle of cilia; in an inner circle was a row of locomotive organs, and the centre exhibited vacuoles constantly opening and shutting. This creature, as before explained, is often found as a parasite upon the polyps. On one occasion a glimpse was caught of a Rotifer similar in shape to the common wheel animalcule, but with a yellow inside. Possibly it was the object so beautifully delineated by Mr. Gosse, in his "Tenby," and described as the "Yellow Philodine," but this must remain in doubt, as it managed to escape before it could be secured.

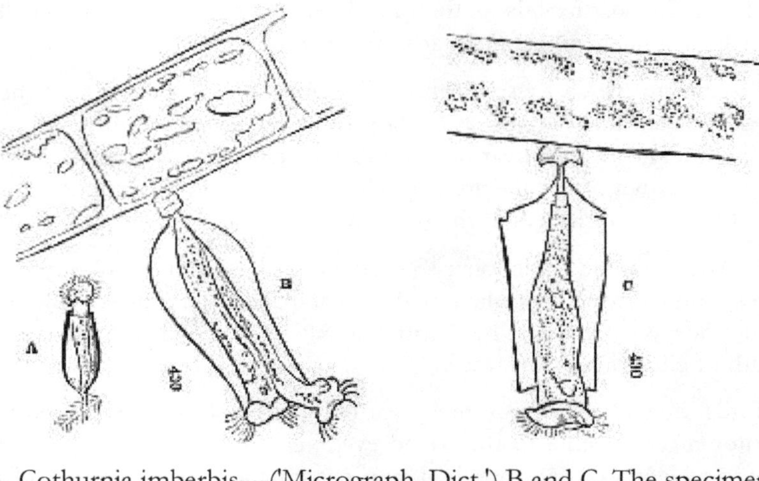

A. Cothurnia imberbis—('Micrograph. Dict.') B and C. The specimens described in text. The figures give the linear magnification.

By the 18th of the month the Vorticellids were much more plentiful, and their changes easily watched; many left their stalks while under the microscope, after which some rushed about like animated and demented hats, others twirled round like tee-to-tums, while others took a rest before commencing their wild career. But the common Vorticellæ were not the only or the most interesting representations of their charming order, for upon some threads of conferva were descried several elegant crystal vases standing upon short foot-stalks, and containing little creatures that jumped up and down like "Jack in the box." These were so minute, that a power of four hundred and thirty linear was advantageously brought to bear upon them. When elongated their bodies were somewhat pear-shaped, but more slender, and variegated with vacuoles and particles of food. The mouths resembled those of Vorticellæ, and put forth circles of vibrating cilia. They were easily alarmed, when the cilia were retracted, and down they sank to the bottom of their vases, quickly to rise again. In one bottle there were two living in friendly juxtaposition. This was not a case of matrimonial felicity, nor of Siamese twins, but of *fission*, or reproduction by division. The original inhabitant of the tube finding himself too fat, or impelled by causes we do not understand, quietly divided himself in two, and as the house was big enough, no enlargement was required. How many stout puffy gentlemen must envy this process; how convenient to have two thin lively specimens of humanity made out of one too obese for locomotion. Man is, however, sometimes the victim of his superior organization, and no process of "fission" can make the lusty lean.[9]

[9] Balbiani in his 'Recherches sur les Phénomènes Sexuels des Infusoires,' speaks of the Vorticellids as the only Infusoria dividing longitudinally. In other species such appearances arise from conjunction.

The bottles in which these creatures live, in happy ignorance that they are called by so crackjaw a name as *Cothurnia imberbis*, were described as *Carapaces* by Ehrenberg, but they bear no resemblance to the shell of a turtle or crab. They are thrown off by the animals who preserve no other connection with them than the attachment at the bottom.

The Micrographic Dictionary describes the family Ophrydina as corresponding to Vorticellina with a carapace. Stein places them with Vorticellids, &c., amongst his Peritricha, which are characterised by a spiral wreath of cilia round the mouth.

Towards the end of the month a great number of black pear-shaped bodies (Stentor niger), visible to the naked eye, were conspicuous in some water from the Kentish Town ponds. Upon examination they were found to be filled with granules that were red by reflected, and purple by transmitted light. Each one had a spiral wreath of cilia, with a mouth situated like those of the stentors, hereafter to be described, but none of them became stationary, and in a few days they all disappeared. Stein divides Ehrenberg's Stentor igneus from S. niger; the creature described seems to have agreed with Stein's *igneus*, which he describes as having blood-red lilac, cinnabar, or brown-red pigment particles, and as much smaller than his S. niger. In the same water were specimens of that singular Rotifer, the *Salpina*, about 1—150" long, and furnished with a *lorica*, or carapace, resembling a three-sided glass box, closed below, and slightly open along the back. At the top of this box were four, and at the bottom three, points or horns, and the creature had one eye and a forked tail. Keeping him company was another little Rotifer, named after its appearance, *Monocerca rattus*, the 'One-tailed Rat.' This little animal had green matter in its stomach, which was in constant commotion. I ought to have observed that the Salpina repeatedly thrust out its gizzard, and used it as an external mouth. In the annexed sketch the Salpina is seen in a position that displays the dorsal opening of the carapace. Its three-cornered shape is only shown by a side view.

Here we close a brief account of what March winds brought in their train. The next chapter will show the good fortune that attended April showers.

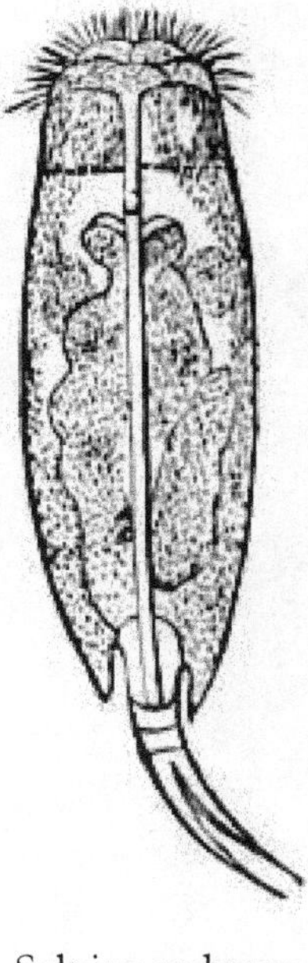

Salpina redunca.

CHAPTER V.

APRIL.

The beautiful Floscule—Mode of seeking for Tubicolor Rotifers—Mode of illuminating the Floscule—Difficulty of seeing the transparent tube—Protrusion of long hairs—Lobes—Gizzard—Hairy lobes of Floscule not rotatory organs—Glass troughs—Their construction and use—Movement of globules in lobes of Floscule—Chætonotus larus—Its mode of swimming—Coleps hirtus—Devourer of dead Entomostraca—Dead Rotifer and Vibriones—Theories of fermentation and putrefaction—Euplotes and Stylonichia—Fecundity of Stylonichia.

FEW living creatures deserve so well the appellation of "beautiful" as the *Floscularia ornata*, or Beautiful Floscule, although to contemplate a motionless and uncoloured portrait, one would imagine that it exhibited no graces of either colour or form. Mr. Gosse has, however, done it justice, and the drawing in his "Tenby" is executed with that rare combination of scientific accuracy and artistic skill, for which the productions of his pencil are renowned.

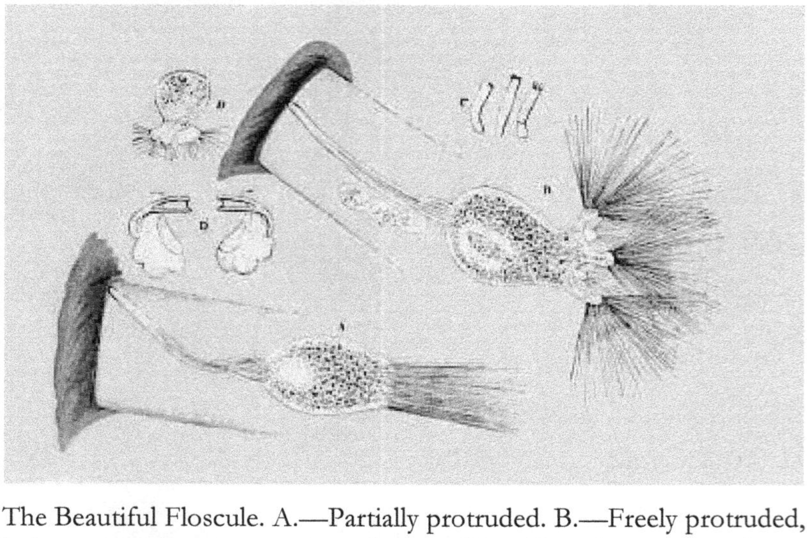

The Beautiful Floscule. A.—Partially protruded. B.—Freely protruded, with three eggs. C.—Appearance of young. D.—Floscule seventeen hours old. D'.—Jaws of Floscule, as figured by Mr. Gosse.

Probably the sketches in several works of authority representing the long cilia as short bristles, are merely copies from old drawings, from objects imperfectly seen under indifferent microscopes, and before the refinements of illumination were understood. Be this as it may, any reader will be fortunate if on an April, or any other morning, he or she effects the capture of one of these exquisite objects, although the first impression may not equal previous expectations, as the delicacy of the organism is not disclosed by a mode of using the light which answers well enough for the common infusoria.

When the Floscules, or other tubicolar Rotifers are specially sought for, the best way is to proceed to a pond where slender-leaved water-plants grow, and to examine a few branches at a time in a phial of water with a pocket-lens. They are all large enough to be discerned, if present, in this manner, and as soon as one is found, others may be expected, either in the same or in adjacent parts of the pond, for they are gregarious in their habits. With many, however, the first finding of a Floscule will be an accident, as was the case last April, when a small piece of myriophyllum was placed in the live-box, and looked over to see what it might contain. The first glimpse revealed an egg-shaped object, of a brownish tint, stretching itself upon a stalk, and showing some symptoms of hairs or cilia at its head. This was enough to indicate the nature of the creature, and to show the necessity for a careful management of the light, which being adjusted obliquely, gave quite a new character to the scene. The dirty brown hue disappeared, and was replaced by brilliant colours; while the hairs, instead of appearing few and short, were found to be extremely numerous, very long, and glistening like delicate threads of spun glass.

Knowing that the Floscules live in transparent gelatinous tubes, such an object was carefully looked for, but in this instance, as is not uncommon, it was perfectly free from extraneous matter, and possessed nearly the same refractive power as the water, so that displaying it to advantage required some little trouble in the way of careful focusing, and many experiments as to the best angle at which the mirror should be turned to direct the light. When all was accomplished, it was seen that the Floscule had her abode in a clear transparent cylinder, like a thin confectioner's jar, which she did not touch except at the bottom, to which her foot was attached. Lying aside her in the bottle were three large eggs, and the slightest shock given to the table, induced her to draw back in evident alarm. Immediately afterwards she slowly protruded a dense bunch of the fine long hairs, which quivered in the light, and shone with a delicate bluish-green lustre, here and there varied by opaline tints.

The hairs were thrust out in a mass, somewhat after the mode in which the old-fashioned telescope hearth-brooms were made to put forth their bristles.

As soon as they were completely everted, together with the upper portion of the Floscule, six lobes gradually separated, causing the hairs to fall on all sides in a graceful shower, and when the process was complete, they remained perfectly motionless, in six hollow fan-shaped tufts, one being attached to each lobe. Some internal ciliary action, quite distinct from the hairs, and which has never been precisely understood, caused gentle currents to flow towards the mouth in the middle of the lobes, and from the motion of the gizzard, imperfectly seen through the integument, and from the rapid filling of the stomach with particles of all hues, it was plain that captivity had not destroyed the Floscule's appetite, and that the drop of water in the live-box contained a good supply of food.

Sometimes the particles swallowed were too small to be discerned, although their aggregate effect was visible; but often a monad or larger object was ingulfed, but without any ciliary action being visible to account for the journey they were evidently compelled to perform. The long hairs took no part whatever in the foraging process, and as they do not either provide victuals or minister to locomotion, they are clearly not, as was supposed by early observers, representatives of the "wheels," which the ordinary Rotifers present. Neither can the cylindrical jar or bottle be justly deemed to occupy the position of the lorica, or carapace which we have before described. The general structure of the creature and the nature of its gizzard distinctly marked it out as a member of the family we call "Rotifers," but the absence of anything like "wheels" proves that those organs are not essential characteristics of this class.

Noticeable currents are not always produced when the mouth of this Floscule is fully expanded. On one occasion, one having five lobes was discovered standing at such an angle in a glass trough that the aperture could be looked down into. The position rendered it impossible to use a higher power than about two hundred linear, but with this, and the employment of carmine, nothing like a vortex was seen during a whole evening, although a less power was sufficient to show the ciliary whirlpools made by small specimens of *Epistylis* and *Vaginicola*, which were in the small vessel. The density of the integument was unfavorable to viewing the action of the gizzard, but it could be indistinctly perceived. The contractions and subsequent expansions of the cup, formed by the upper part of the creature, may be one way in which its food is drawn in, but there is no doubt it can produce currents when it thinks proper. Sometimes animalcules in the vicinity of Floscules whirl about as if under the influence of such currents. Some may be seen to enter the space between the lobes, swim about inside, and then get out again, while every now and then one will be sucked in too far for retreat.

Above the gizzard in the Horned Floscule,[10] I have seen an appearance as if a membrane or curtain was waving to and fro, while another was kept in a fixed perpendicular position. Mr. Gosse, speaking of this genus, observes "that the whole of the upper part of the body is lined with a sensitive, contractile, partially opaque membrane, which a little below the disk recedes from the walls of the body, and forms a diaphragm, with a highly contractile and versatile central orifice. At some distance lower down another diaphragm occurs, and the ample chamber thus enclosed forms a kind of *crop*, or receptacle for the captured prey."

[10] The Horned Floscules (*F. cornuta*) which I have found, and which bred in a glass jar, were not so large as those described by Mr. Dobie, as quoted in 'Pritchard's Infusoria.' Mr. Dobie's specimens were 1—40" when extended; mine about half that size, five-lobed, and with a long slender proboscis, standing in a wavy line outside one lobe. Mr. Dobie also describes an *F. campanulata*, with five flattened lobes. The 'Micrographic Dictionary' pronounces these two species "doubtfully distinct." I have three or four times met with a variety of *F. ornata*, in which one lobe was much enlarged and flattened, but they had no proboscis. In what I take for *F. cornuta*, the horn or proboscis has sometimes been a conspicuous object, and at others so fine and transparent as to be only visible in certain lights.

"From the ventral side of the ample crop that precedes the stomach, there springs in *F. ornata* a perpendicular membrane or veil, partly extending across the cavity. This is free, except at the vertical edge, by which it is attached to the side of the chamber, and being ample and of great delicacy, it continually floats and waves from side to side. At the bottom of this *veil*, but on the dorsal side, are placed the jaws, consisting of a pair of curved, unjointed, but free *mallei*, with a membranous process beneath each."

The Beautiful Floscule could always be made to repeat the process of retreating into her den, and coming out again to spread her elegant plumes before our eyes, by giving the table a smart knock, and her colours and structure were well exhibited by the dark-ground illumination, which has been explained in a previous page.

An object like this should be watched at intervals for hours and even days, especially if the eggs are nearly ready to give up their infantile contents. This was the case with the specimen described, and after a few hours a young Floscule escaped, looking very much like a clumsy little grub. After a few awkward wriggles the new-born baby became more quiet, and on looking at it again at the expiration of seventeen hours, it had developed into the shape of a miniature plum-pudding, with five or six tiny lobes expanding their tufts of slender hair. Unfortunately its further proceedings were not seen, or it would have been interesting to note the growth of the foot, and the formation of the gelatinous tube, which is probably thrown off in rings.

To view the details of the structure of a Floscule, it must be placed in a live-box or compressorium, and if specimens are scarce, they should not be allowed to remain in the limited quantity of water those contrivances hold, after the observations are concluded, but should be carefully removed, and placed in a little vial, such as homœopathists use for their medicine. By such means an individual may be kept alive for many days. It is also interesting to place a little branch of the plant occupied by Floscules or similar creatures, in a glass trough, where they may be made quite at home, and their proceedings agreeably watched by a one-inch or two-thirds power. These troughs,[11] which can be obtained of the optician, should be of plate glass, about three inches long, nearly the same height, and about half an inch wide. If narrower, or much taller, they will not stand, which is a great inconvenience. The pieces of glass are stuck together with marine glue, and a very simple contrivance enables the plants or other objects to be pressed near the front, and thus brought into better view. A strip of glass, rather narrower than the width of the trough, is dropped into it, and allowed to fall to the bottom. Then a piece of glass rather shorter than the trough, and rather higher than its front side, is placed so as to slope from the front of the bottom towards the back at the top. The piece of glass first dropped in keeps it in the right position, and the trough is thus made into a V-shaped vessel, wide at the top and gradually narrowing. Any object then placed in it will fall till it fits some part of the V, where it will remain for observation. A small wedge of cork enables the moveable piece of glass to be thrown forwards, until it assumes any angle, or is brought parallel to the front of the trough.

[11] The shallow cells with thin sliding covers devised by Mr. Curteis (of Baker's), are still more convenient when no pressure is required, and the objects are small. When not under the microscope they can be kept full of water by immersion in a tumbler.

A power of five or six hundred diameters generally enables a movement of small globules to be seen at the extremity of the lobes of the Floscule, and the gizzard may be made plain by dissolving the rest of the creature in a drop of solution of caustic potash. It also becomes more visible as the supply of food falls short. Mr. Gosse describes the body as "lined with a yellowish vascular membrane," and young specimens exhibit two red eyes, which may or may not be found in adults. When these eyes of Rotifers are not readily conspicuous, they must be sought for by opaque illumination, or by the dark-ground method which, especially with the parabola, is successful in bringing them out.

Naturalists, and possibly the specimens also, do not always agree in the number of lobes assigned to the "Beautiful Floscule," and although it is easy enough to count them in *some* positions, the observer may have to exercise a good deal of patience before he is certain whether they are five or six. For a long evening only five could be discerned in the specimen now described, but the next night six were apparent without difficulty or doubt. The hairs also will not appear anything like their true length or number, unless the object-glass is good, and great care is taken not to obscure them by a blaze of ill-directed light.

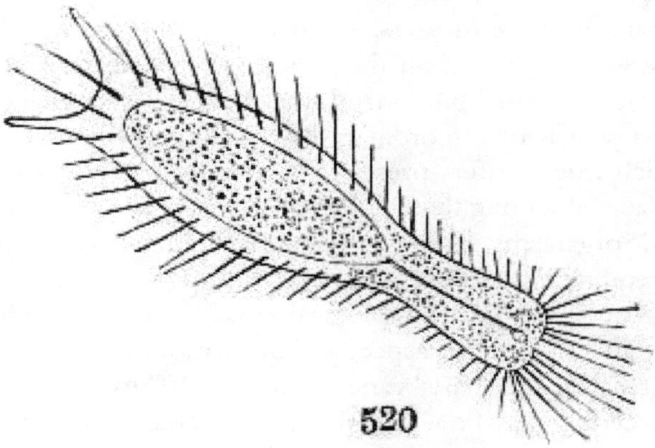

520

Chætonotus larus (swimming).

After the Floscules had been sufficiently admired and put aside, for observations to be repeated on future occasions, a Rotifer attracted attention by his merry-andrew pranks, throwing himself in all directions by means of two long and extremely mobile toes attached to his tail-foot. Then came a creature swimming like an otter, thrusting his head about on all sides, and looking much more intelligent than most of his compeers of the pond. Looked at vertically, he was somewhat slipper-shaped, the rounded heel forming his head, then narrowing to a waist, and expanding towards the other end, which projected in a fork. All round him were long cilia, which were conspicuous near the head, and a fine line indicated the passage from his mouth to the stomach, which seemed full of granular matter. Presently he took to crawling, or rather running, over a thread of conferva, and then his back was elegantly arched, and his cilia stood erect like the quills of a porcupine. This was the *Chætonotus larus*.

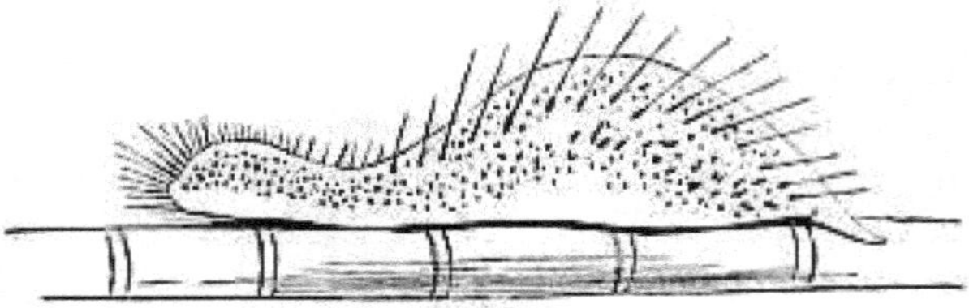

520

Chætonotus larus (crawling).

In Pritchard's "Infusoria," the views of those writers are followed who rank this animal amongst the Rotifers, and place it in the family *Icthidina*. To help out this theory, the cilia upon the ventral surface are imagined to form a "band-like rotary organ;" but in truth they bear no resemblance whatever to the so-called wheels of the ordinary Rotifers, nor is there anything like the gizzard which true Rotifers present. Ehrenberg treated it as a Rotifer, and Dujardin placed it among the Infusoria, in a particular class, comprehending symmetrical organisms. The 'Microscopic Dictionary' remarks that its "structure requires further investigation,"[12] and while the learned decide all the intricate questions of its zoological rank, the ordinary observer will be pleased to watch its singular aspect and lively motions. Its size, according to the 'Micrographic Dictionary,' varies from 1—710" to 1—220", and while its general proceeding may be watched with an inch or two-thirds object-glass, and the second eye-piece, a power of five hundred linear (obtained by a quarter or a fifth) is required to make out the details of its structure. If placed in a live-box with threads of conferva, and a little decayed vegetation, it may be observed to group about among them, and shake them like a dog.

[12] See a valuable paper by Mr. Gosse, "History of the Hairy-backed Animalcules," 'Intellectual Observer,' vol. v, p. 387, in which the known species are described and reasons given for following Vogt and ranging them with the Turbellarian worms.

We have said that water-fleas were among the inhabitants of a bottle filled at the pond, and as they go the way of all flesh, it is common to find some odd-looking animalcules ready to devour their mortal remains. These are creatures shaped like beer-barrels, upon short legs, and which swim with a tubby rolling gait. Looking at one of these little tubs lengthwise, a number of lines are seen, as though the edge of each stave projected a little above the general level, and transverse markings are also apparent, which may be compared to hoops. This is the *Coleps hirtus*, which differs from the usual type of Infusoria, by being symmetrical, that is, divisible into two equal and similar halves. The dimensions of this species vary from 1—570 to 1—430, and its colour varies from white to brown. It has been observed to increase by transverse self-division, and has two orifices, one at each end, for receiving food and ejecting the remains. It often requires some little trouble to get a good view of the cilia, which are arranged in transverse and longitudinal rows. A power of one hundred and fifty linear is convenient for viewing it in motion, but when quiet under pressure, one of five or six hundred may be used with advantage.

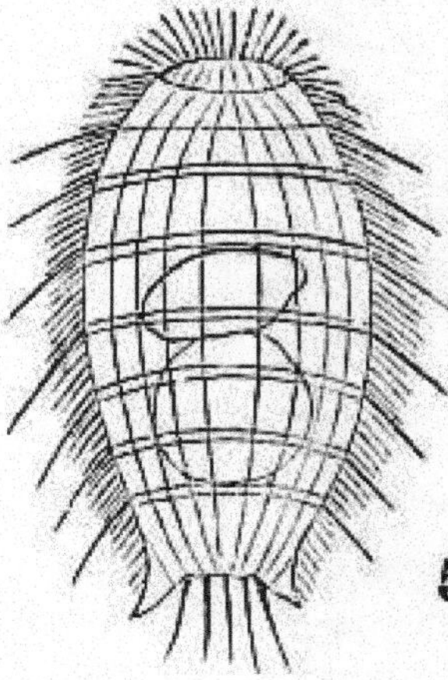

565

Coleps hirtus.

Among the rubbish at the bottom of the bottle, in which the coleps was found, was a minute dead Rotifer, the flesh of which was fast disappearing, but upon being examined with a power of nine hundred and sixty diameters, it was observed to swarm with extremely minute *vibriones*, the largest only appearing under that immense magnification like chains of bluish-green globules, not bigger than the heads of minikin pins, while the smallest were known by a worm-like wriggling, although their structure could not be defined. These *vibriones* are probably members of the vegetable world, and they always appear when animal matter undergoes putrefaction.

M. Pasteur has brought forward elaborate experiments to show that the development of the yeast plant is an act correlative to alcoholic fermentation, and in like manner the growth of *vibriones* may stand in correlation to putrefactive decomposition.

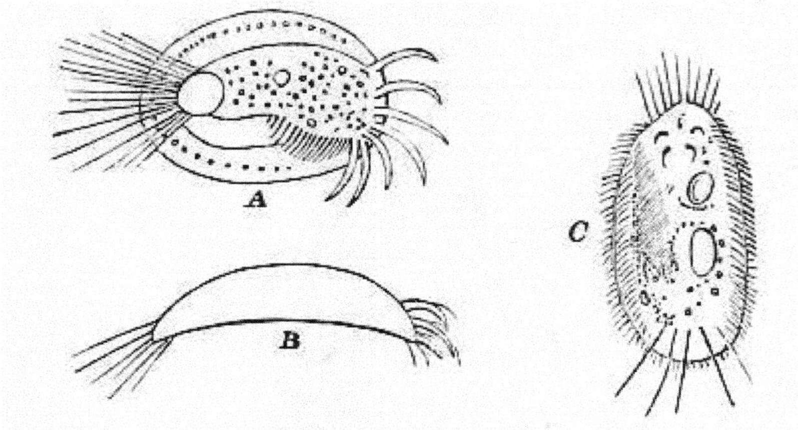

A, Euplotes (patella); B, side view of ditto; C, stylonichia.

Ehrenberg considered them animals, and fancied he detected in them a plurality of stomachs; but the vegetable theory is the more probable, at any rate of the species under our notice, which is often seen, though not always so minute.

At this time two interesting animalcules were very plentiful—the *Euplotes patella*, and *Stylonichia*, both remarkable as exhibiting an advance in organization, which approximates them to the higher animals. In addition to cilia they possess *styles*, which take the place of the limbs of more elaborately-constructed creatures, and give a variety to their means of locomotion. The *Euplotes* is furnished with an oval carapace covering the upper surface, which in different individuals, and probably at different ages, exhibits slightly varied markings round its margin, which in the specimen drawn above consisted of dots. They can run, climb, or swim, and exemplify a singular habit which several of the infusoria possess, that of moving for a little time in one direction, and then suddenly, and without any apparent cause, reversing it. If the reader is fond of learned appellations, he can call this *diastrophy*, but we do not know that he will be any the wiser for it.

The Stylonichia are oval animalcules, surrounded by cilia, and having moreover a collection of styles, both straight and curved, the latter called *uncini*, or little hooks. They swim steadily on, and then dart back, but not so far as they have advanced, and may be seen to keep up this fidgety motion by the hour together. Pritchard tells us Ehrenberg found that a single animalcule lived nine days; during the first twenty-four hours it was developed by transverse self-division into three animals; these in twenty-four hours formed two each in the same manner, so that by self-division only (without ova), these animalcules increased three or four-fold in twenty-four hours, and *may thus produce a million* from a single *animalcule in ten days*. Such are the amazing powers of reproduction conferred upon these humble creatures, powers which are fully employed when the surrounding circumstances are favorable, and which, in the aggregate, change the condition of large masses of matter, and bring within the circle of life millions upon millions of particles every minute of the day.

CHAPTER VI.

MAY.

Floscularia cornuta—Euchlanis triquetra—Melicerta ringens—its powers as brickmaker, architect, and mason—Mode of viewing the Melicerta—Use of glass-cell—Habits of Melicerta—Curious Attitudes—Leave their tubes at death—Carchesium—Epistylis—Their elegant tree forms—A Parasitic Epistylis like the "Old Man of the Sea"—Halteria and its Leaps—Aspidisca Lynceus.

MAY, the first of summer months, and of old famous for floral games, which found their latest patrons in the chimney-sweeps of London, is a good time for the microscopist among the ponds, for the increase of warmth and heat favours both animal and vegetable life, and so we found as we carried home some tops of myriophyllum, and soon discovered a colony of tubicolor rotifers among the tiny branches. They proved to be Floscules, generally resembling the *F. ornata*, described in a previous page, but having a long slender proboscis hanging like a loose ringlet down one side. The cilia or hairs were not so long as in the Beautiful Floscules we had before obtained, nor was their manner of opening so elegant; but they were, nevertheless, objects of great interest, and were probably specimens of the *Floscularia cornuta*. A swimming rotifer in a carapace somewhat fiddle-shaped, with one eye in its forehead, and a two-pronged tail sticking out behind (the *Euchlanis triquetra*), also served to occupy attention; but a further search among the myriophyllum revealed more treasures of the tube-dwelling kind. These were specimens of that highly curious Rotifer, the *Melicerta ringens*, who, not content with dwelling, like the Floscules, in a gelatinous bottle, is at once brickmaker, mason, and architect, and fabricates as pretty a tower as it is easy to conceive. The creature itself stands upon a retractile foot-stalk, and thrusts out above its battlements a large head, with four leaf-like expansions surrounded by cilia. Between the lower lobes, or leaves, the gizzard is seen grinding away, and above it is an organ, not always displayed, and of which Mr. Gosse was fortunate enough to discover the use. This eminent naturalist likens it to the circular ventilator sometimes inserted in windows, and he found it was the machine for making the yellow ornamental bricks of which the tower is composed. Pellet by pellet, or brick by brick, does the Melicerta build her house, which widens gradually from the foundation to the summit, and every layer is placed with admirable regularity.

In order to obtain the materials for her brickmaking the Melicerta must have the power of modifying the direction of the ciliary currents, so as to throw a stream of small particles into the mould, which is a muscular organ, and capable of secreting a waterproof cement, by which they are fastened together. The result is, not to produce anything like the tubes made by the caddis-worms out of grains of sand, but entirely to change the appearance of the materials employed. All large particles are rejected, and only those retained which will form a homogeneous pulp with the viscid secretion; and when the process is complete the head of the creature is bent down, and the pellet deposited in its appropriate place. Each pellet appears originally to possess a more or less conical figure, but when they are squeezed together to make a compact wall they all tend to a hexagonal form, by which they are able to touch at all points, and any holes or interstices are avoided.

According to Professor Williamson the young Melicerta commences her house by secreting "a thin hyaline cylinder," and the first row of pellets are deposited, not at the base as would be expected, but in a ring about the middle of the tube. "At first new additions are made to both extremities of the enlarging ring; but the jerking constrictions of the animal at length force the caudal end of the cylinder down upon the leaf, to which it becomes securely cemented by the same viscous secretion as causes the little spheres to cohere."

Round the margins of the lobes or expansions may be seen delicate threads towards which others radiate; these are thought by Mr. Gosse to be portions of a nervous system, and two calcars or feelers serve as organs of relation. The young Melicertas are likewise furnished with a pair of eyes, which are probably rudimentary, and disappear as they grow up.

The Melicerta tubes, being large enough to be visible to the naked eye, are easily crushed in the live-box, and to avoid this, they are conveniently viewed in a shallow glass cell, covered up as before described. By occasionally changing the water one may be kept for days in the same cell, and will reward the pains by frequently exposing its flower-like head. Usually the horns or feelers come out first, and then a lump of flesh. After this, if all seems right, the wheels appear, and make a fine whirlpool, as may be readily seen by the use of a little indigo or carmine.

The Melicerta is, however, an awkward object to undertake to show to our friends, for as they knock at the door she is apt to turn sulky, and when once in this mood it is impossible to say when her fair form will reappear. At times the head is wagged about in all directions with considerable vehemence, playing singular antics, and distorting her lobes so as to exhibit a Punch and Judy profile. When these creatures die they leave their tubes, which are often found empty in the ponds they frequent. The Melicertas are conveniently viewed with a power of from sixty to one hundred linear, and a colony of them may be kept alive for some weeks in a glass jar or tank.

Among the remainder of my tiny captives were two beautiful members of the Vorticella family, *Epistylis* and *Carchesium*. The reader will remember that in the Vorticella previously described, the bells stood upon stalks that were very flexible, and retractile by means of a muscle running down their length. The *Epistylis* is, as its name imports, the dweller on a *pillar*. The stem is stiff, or only slightly flexible, and has no apparatus by which it can be drawn down. The specimen mentioned stood like a palm-tree, and the large oval bells drooped elegantly on all sides, as its portrait will show. At times they nodded with a rapid jerk.

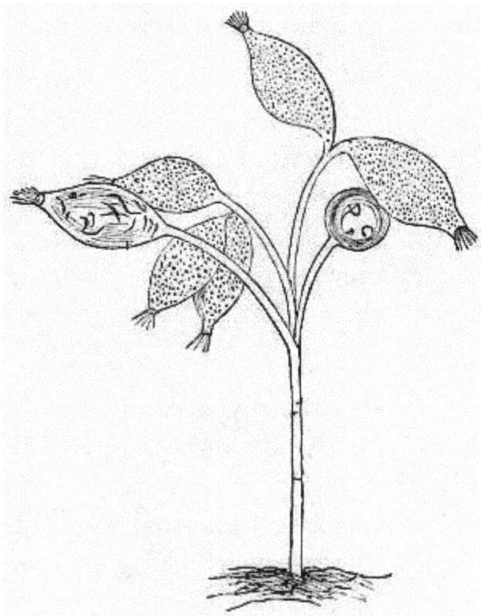

Epistylis.

The *Carchesium* differs from the common *Vorticella*, by branching like a tree, but the stems are all retractile, although the trunk seldom exercises the power. A group of these creatures presents a spectacle of extraordinary beauty—it looks like a tree from fairy-land, in which every leaf has a sentient life. In general structure the bells of the *Epistylis* and the *Carchesium* resemble the common *Vorticella*, and like them may be seen with a power of about one hundred linear for general effect, and with a higher one for the examination of special points. Pritchard notices three species of *Carchesium*, and eighteen of *Epistylis*;[13] some of which it is to be hoped will turn out to be only varieties.

[13] An interesting *Epistylis*, called *Digitalis*, from its bells resembling fox-glove flowers in shape, occurs as a parasite upon the *Cyclops quadricornis*, a very common entomostracan in fresh-water ponds. At this moment I have

a beautiful specimen, branching like a bushy tree, and attached to the tail of a *Cyclops*, who can scarcely move under his burden, which is like Sinbad's "Old Man of the Sea." (See illustration above.)

Towards the end of this month rotifers abounded, and polyps were plentiful. Among the rotifers was one about a two-hundredth of an inch long, protected by a carapace, and having a tail terminating in a single style, hence called "Monostyle." There is perhaps no class of creatures that present so many curious and unexpected forms as the rotifers; and although we have noticed a good many, there are far more that remain to be found and described.

The water in which the preceding animals dwelt was enlivened by the jumps of the *Halteria*, a little globe surrounded by long fine cilia, with which its movements were effected; and its companion was the *Aspidisca lynceus*, an oval animalcule, having a distinct cilia or *lorica*, and furnished, in addition to cilia, with bristles, which enable it to walk and climb as well as swim.

There were also some eggs of rotifers attached to the water plants, in which motion could be descried at intervals, and a little red eye observed. These eggs are always large in proportion to the creatures that lay them, and if they escape being devoured by enemies, may be watched until their contents step forth.

In this, as in other months, omission is made of creatures that have already come under notice, or our list would assume larger dimensions.

CHAPTER VII.

JUNE AND JULY.

Lindia Torulosa—Œcistes Crystallinus—A professor of deportment on stilts—Philodina—Changes of form and habits—Structure of Gizzard in Philodina family—Mr. Gosse's description—Motions of Rotifers—Indications of a will—Remarks on the motions of lower creatures—Various theories—Possibility of reason—Reflex actions Brain of insects—Consensual actions—Applications of physiological reasoning to the movements of Rotifers and Animalcules.

APRESSURE of other occupations prevented full use being made of June and July, nor was the weather at all propitious. For this reason the microscopic doings of these two months are recorded in one chapter.

As usual the Kentish Town ponds were productive of objects, and among them were several rotifers not found in the previous months. The first of these was a very small worm-like thing, with one eye, a tuft of cilia about the mouth, and two toes at the tail end. Had it not been for the jaws, which were working like fingers thrust against each other, and which were unmistakably of the rotifer pattern, the animal might have been supposed to belong to some other class. According to the 'Micrographic Dictionary,' the *Lindia torulosa* is 1—75" long, but this specimen was only about 1—200". It was possibly very young, and did not thrust out its cilia in two distinct tufts, as Cohn describes, although it may have had the power of doing so. At times it sprang quickly backwards and forwards, bringing its head where its tail was before. This object required for its comfortable elucidation a power of about six hundred linear.

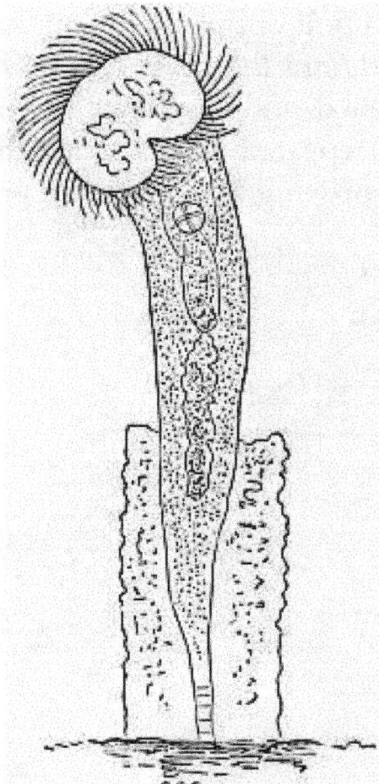

240

Œcistes crystallinus.

Among the common water-plants, which are worth examining as the probable abodes of rotifers or infusoria, is the pretty little thing called "star-weed," some of which was obtained from the last-mentioned ponds, and on examination yielded a specimen of a tube-dwelling rotifer, the *Œcistes crystallinus*, which, although less beautiful than the Floscules or the Melicerta, is, nevertheless, a pretty and interesting object. In this instance a little rough dirty tube, about 1—70" long, was observed to contain an animal capable of rising up and expanding a round mouth garnished with a wreath of cilia; while a little below, the indefatigable and characteristic gizzard of the tribe was in full play. A power of two hundred and forty linear sufficed to afford a good view, and it was seen that a long, irregular, conical body was supported upon a short wrinkled stalk. The usual drawings represent this creature with a short bell-shaped body upon a very long slender pedicle. Possibly this one might have been able to show himself under this guise, but he did not attempt it; his appearance being always pretty much as described, which made the foot shorter and the body longer than the measurements which naturalists have given, and according to which the whole creature is 1—36" long, although the body is only 1—140". The tube of the *Œcistes* is called a "lorica," or carapace; but it has in truth no right whatever to the appellation.

Another strange rotifer, of whose name I am uncertain, had an ovalish oblong body, and a pair of legs like compasses, twice as long as himself. His antics were those of a posture-master, or "Professor of Deportment" on stilts. Sometimes he stood bolt upright, bringing his legs close together; then they were jauntily crossed, and the body carried horizontally; then the two legs would be slightly opened, and the body thrown exactly at right-angles to them. These antics were repeated all the while the observation lasted, and had a very funny effect in proving that drollery is practised, if not understood, in the rotatorial world.

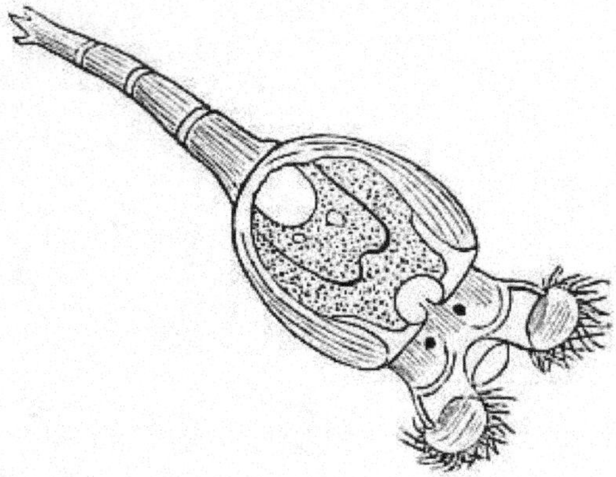

Philodina (swimming).

Another kind of rotifer was abundant—the *Philodina*, which belongs to the same family as the common wheel-bearer, namely, the *Philodinæa*. The *Philodina* is a good deal like the common wheel-bearer, or *Rotifer vulgaris*, but is usually of a stouter build, and carries his eyes in a different place. In the common rotifer these organs are situated on the proboscis, while those of the Philodina are lower, and said to be "cervical." The changes of form in this rotifer are still more remarkable than in the common wheel-bearer. When resting it resembles a pear-shaped purse, puckered in at the mouth. Then it thrusts out its tail-foot, swells its body to an oval globe, protrudes its feeler, and slightly exposes a row of cilia. After this two distinct wheels are everted, and as their cilia whirl and spin, the animal is swiftly rowed along, until it thinks proper to moor itself fast by the tail-foot, and employ all its ciliary power in causing currents to converge towards its throat. When it pleases it can elongate the body, till it becomes vermiform, and it walks like the common rotifer, by curving its back, and bringing its nose and its tail in contact with the ground.

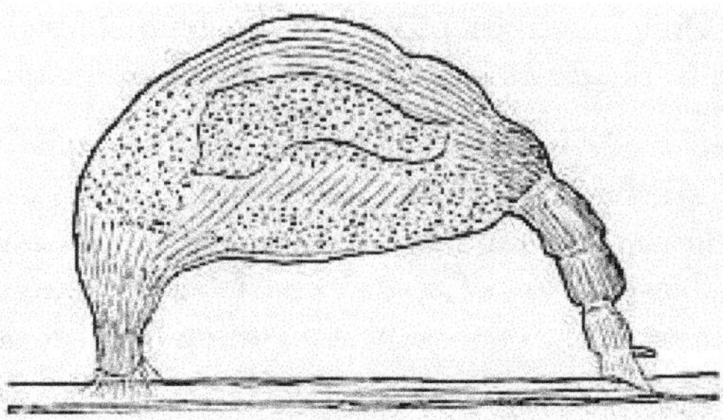

Philodina (crawling).

The gizzard of this family (*Philodinæa*) presents a considerable deviation from the perfect form exhibited by the *Brachions*. According to Mr. Gosse, "The *mallei* and the *incus* (terms already explained) are soldered together into two subquadrantic-globular masses, which appear to be muscular, but invested with a solid integument. The *manubria* (handles) may still be recognised in a vertical aspect as three loops, of which the central one is chiefly developed, and in a vertical aspect as a translucent reniform (kidney-shaped) globe." These descriptions are not easy to understand, not from any want of clearness or precision in the words employed, but from the complicated character of the organ, and its very different appearance under different aspects. To make the matter more intelligible, Mr. Gosse adds, "the structure and action of an apparatus of this type may be made more clear by a homely illustration. Suppose an apple to be divided longitudinally, leaving the stalk attached to one half. Let this now be split again longitudinally so far as the stalk, but not actually separating either portion from it. Draw the two portions slightly apart, and lay them down on their rounded surfaces. They now represent the quadrantic masses in repose, the stalk being the fulcrum, and the upper surfaces being crossed by the teeth. By the contraction of the muscles, of which they are composed, the two segments are made to turn upon their long axis, until the points of the teeth are brought into contact, and the toothed surfaces rise and approach each other. The lower edges do not, however, separate as the upper edges approach, but the form of the mass alters, becoming more lenticular, so that when the toothed surfaces are brought into their closest approximation, the outline has a subcircular figure. It is on account of this change of form that I presume the masses themselves to be partially composed of muscle."

These remarks, although specially made of the *Rotifer macrurus*, are in the main applicable to all the Philodinas, but the student must not expect to understand any of the complicated gizzards of the rotifers without repeated observations, and no small exercise of patience. It is common to call the portions of the Philodine-pattern gizzard "stirrup-shaped," but Mr. Gosse has shown them to be *quadrantic*, that is, shaped like the quarter of a sphere.

As we are not very well off with subjects for description in these two months, we can afford a little time to consider a question that continually arises in the mind, on viewing the movements of animalcules, and especially of any so highly developed as the rotifers, namely, to what extent motions which appear intelligent are really the result of anything like a conscious purpose or will. When any of the lower animals—a bee, for example—acts in precisely the same way as all bees have acted since their proceedings have been observed, we settle the question by the use of the term *instinct*. Those who take the lowest view of insect life, assume that the bee flies because it has wings, but without wishing to use them, and that the nerves exciting them to action are in their turn excited, not by volition, but by some physical stimulus.

The sight or the smell of flowers is thought by the same reasoners to be capable of attracting the insect, which is unconscious of the attraction, while proximity of food stimulates the tongue to make the movements needful for its acquisition, and so forth. The cells, they tell us, are built according to a pattern which the earliest bee was impelled to construct by forces that bear no analogy to human reason and human will, and so originate all the ordinary processes of bee life. Sometimes, however, it happens that man or accident interposes particular obstacles, and forthwith there appears a particular modification of the orthodox plan, calculated to meet the special difficulty. How is this? Does any one of the difficulties which the bee or the ant is able to get over, produce precisely that kind of electrical disturbance, or polar arrangement of nerve particles that is necessary to stimulate the *first* step of the action by which the difficulty is surmounted; and does the new condition thus established stimulate the *second* step, and so forth, or can the bee, within certain limits, really *think*, design, and contrive?

No questions are more difficult of solution; but while protesting against a tendency to undervalue all life below that of man, we must remember we have in our bodies processes going on which are not the result of volition, as when the blood circulates, and its particles arrange themselves in the pattern required to form our tissues and organs, and also that many of our actions belong to the class termed by physiologists, "reflex," that is, the result of external impressions upon the nervous system, in which the *sentient* brain takes no part. Thus when a strong light stimulates the optic nerve, the portion of brain with which it is connected in its turn stimulates the iris to contract the pupil; and it is supposed that after a man has begun to walk, through the exercise of his will, he may continue to walk, by a reflex action; as his feet press the ground they transmit an impression to the spinal cord, and the legs receive a fresh impulse to locomotion, although the mind is completely occupied with other business, and pays no attention to their proceedings.[14]

The ordinary movements of insects appear to be of this character, and to be excited by the ganglia belonging to the segment to which the moving limbs are attached. Thus a centipede will run, after its head has been cut off, and a water-beetle (*Dytiscus*) swam energetically when thrown into water after its brain had been removed.[15]

[14] See Carpenter's 'Manual of Physiology.'

[15] Carpenter's 'Manual of Physiology,' p. 551.

It must not, however, be assumed that the brain of insects has nothing to do with their movements. It is probably the means of co-ordinating or directing them to a common end, and gives rise to what are called *consensual* movements, that is, movements which are accompanied or stimulated by a sensation, although not controlled by a will. In man these actions are frequently exhibited, "as when laughter is provoked by some ludicrous sight or sound, or by the remembrance of such at an unseasonable hour."[16] Sneezing is another instance of a sensation leading to certain motions, without any intervention of the human will.

[16] Ibid., p. 543.

Speaking of these consensual motions, Dr. Carpenter observes, "It is probable, from the strong manifestations of emotion, exhibited by many of the lower animals, that some of the actions which we assemble under the general designation of instinctive are to be referred to this group."

The insect brain is composed of a supra-œsophagal ganglion and infra-œsophagal one. Von Siebold says, the first corresponds to the cerebrum of the vertebrata, and "the second is comparable, perhaps, to the cerebellum or spinal cord."[17] The superior ganglion gives off nerves to the antennæ and eyes, the lower one to the mandibles, &c. So far as is known the insects that exhibit the most intelligence have the largest and best developed brains.

[17] 'Anatomy of Invertebrates,' Burnett's trans.

A special volume would be required for anything like a complete examination of the little which is known on this subject, but these few remarks may assist the microscopic beginner in examining the movements of his subjects, and guard against the error of referring to reason and volition those which are, probably, either the direct result of stimulants applied to the surface (as in nerveless creatures), or the indirect (reflex) result of such stimulants in beings like the rotifers, who have a nervous system; or the result of *sensations*, which excite actions without previously referring the matter to the decision of a will. It must not, however, be too readily assumed that the behaviour of creatures possessing distinct organs is entirely automatic; and we must not forget that even the best physiologists know very little concerning the range of functions which the nervous ganglia of the invertebrata are able to discharge.

CHAPTER VIII.

AUGUST.

Mud coloured by worms—Their retreat at alarm—A country duck-pond—Contents of its scum—Cryptomonads—Their means of locomotion—A Triarthra (three-limbed Rotifer)—The Brachion or Pitcher Rotifer—Its striking form—Enormous gizzard—Ciliary motion inside this creature—Large eye and brain—Powerful tail—Its functions—Eggs.

IN the beginning of this month a pond in the Finchley Road, a little beyond the Highgate Archway, supplied some more specimens of the *Pterodina patina*, described in a previous chapter; but towards the middle of the month a visit to Chipstead, in Surrey, enabled a new region to be explored.

It is always a treat to a Londoner to get down to any of the picturesque parts of Surrey; the trees exhibit a richness of foliage and variety of colour not seen within the regions of metropolitan smoke; the distance glows with the rich purples so much admired in the pictures of Linnel, and the sunsets light up earth and sky with the golden tints he is so well able to reproduce. Probably the warmth of the soil, and the purity of the air, may make Surrey ponds prolific in microscopic life; but of this we do not know enough to make a fair comparison, although our own dips into them were tolerably lucky.

Walking one day down a lane leading towards Reigate, where the trees arched overhead, ferns grew plentifully in the sandy banks, and the sunlight flitted through the branches, and chequered the path, we came to a shallow pond, or great puddle, which crossed the way, and near the edge of the water the eye was struck with patches of crimson colour. On attempting to take up a portion of one of these patches the whole disappeared, although when the disturbance ceased the rich colour again clothed the dingy mud. The appearance was caused by thousands of little worms, belonging to the genus *Tubifex*, not uncommon in such situations, who thrust themselves out to enjoy the light and air, and retreat the moment an alarm is given. Probably both actions belong to the class described in the last chapter, as "reflex;" but it would be interesting to know whether creatures so humble have any sense of fear. These worms will repay observation, but in these pages we eschew all their tribe—unless the rotifers be assigned to them—and take ourselves once more to our especial subjects.

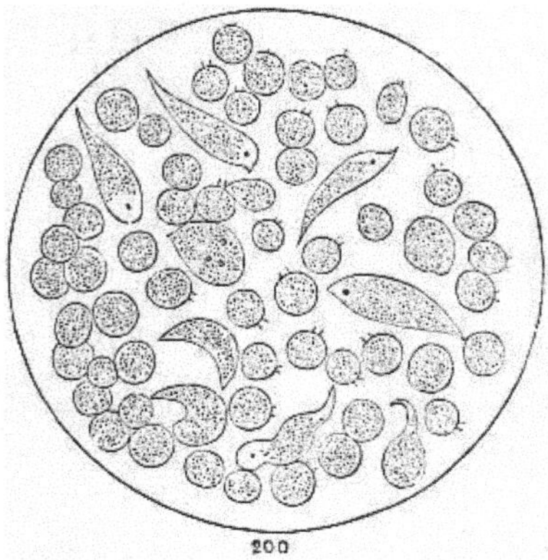

Cryptomonad—Euglena.

Knowing that farm-ponds are usually well stocked with microscopic game, we made a dip into one more especially assigned to ducks, and obtained wondrous little for our pains. We were not, however, discouraged, but made an examination of the circumstances, which determined a particular course of action. Our piece of water was simply a dirty duck-pond, in which no large plants were growing, and which did not even exhibit the little disks of duckweed that are common to such situations. There was, however, on the surface, in parts, an exceedingly fine scum of pale yellow green, and this, armed with a teaspoon, we proceeded to attack. By careful skimming, a small bottle was half-filled with minute organic particles, which were likely to be interesting in themselves, and pretty sure to be the food for something else. A small drop was placed on a tablet of the live-box, flattened out by the application of the cover, and viewed with a power of two hundred linear, which disclosed swarms of brilliant green globes, amongst which were a good sprinkle of minute creatures, like the *Euglenæ* already described, and whose little red eyes contrasted vividly with the prevailing emerald hue.

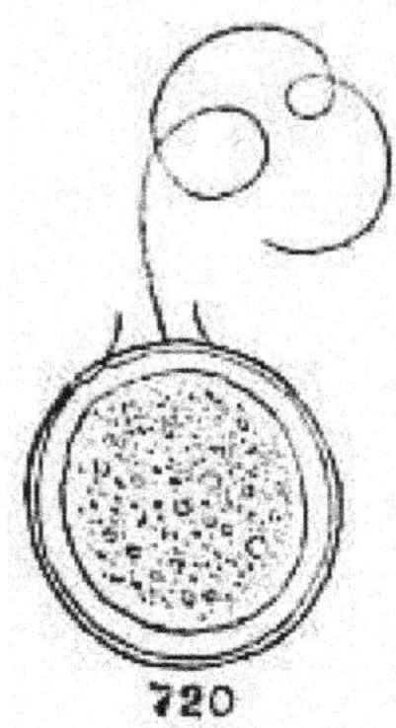

Cryptomonad.

One of the higher infusoria, whose species I could not identify, was devouring them like a porpoise among sprats. It did not, however, exhibit any sense in its hungry career; it moved about in all directions, gulping down what came in its way, but often permitting the escape of the little green things that were almost in its mouth. The little globes rolled and whirled about without the faintest indication of a purpose, and without exhibiting any instrument with which their locomotion was effected. To find out how this was done, a higher power was used, and from their extreme minuteness an amplification of seven hundred and twenty linear was conveniently employed, although a lower one (three or four hundred) disclosed the secret by showing that a little whip was flourished about through the neck, which the lower power revealed. When highly magnified, each little globe was seen to consist of an outer case of a reddish orange colour, which was noticeable on looking at the edges, although in the centre it was transparent enough to show the brilliant green contents, that resembled the chlorophyll, or green colouring matter of plants. From a short neck proceeded the whip-like filament, which was lashed and twisted about in all directions. These little creatures belong to the monad family, but whether they are to be called *Trachelomonads*, or by some other hard name, the learned must decide.

The 'Micrographic Dictionary' puts a note of interrogation to the assertion of some writers that *Trachelomonads* have no necks, and draws some with such an appendage.

Pritchard's last edition is against necks, and whether the necks or no necks are to win, is a mighty question equal at least to the famous controversy, which divided the world into "big and little endians in the matter of breaking eggs."

A discussion of more importance is, whether these *Cryptomonads*—that name will do whatever comes of the neck controversy—are animals or vegetables. Lachmann and Mr. Carter affirm that they have detected a contractile vesicle, which would assimulate them to the animal series, but their general behaviour is vegetable; and the 'Micrographic Dictionary' is in favour of referring them to the *Algæ*—that great family of simple plants, of which the sea-weeds are the most important representatives.

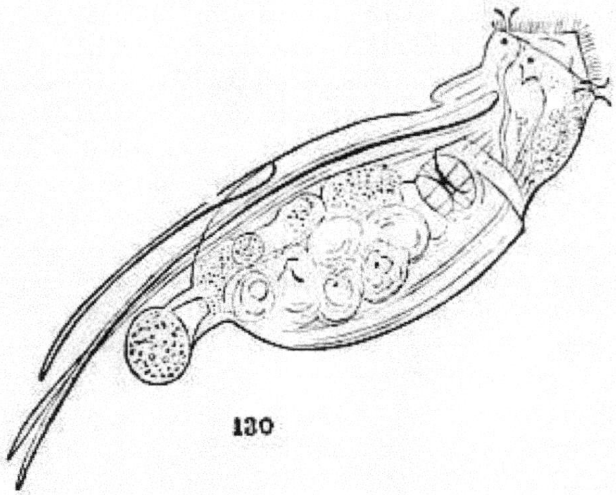

130

Triarthra.

When any of the monads swarm, there are sure to be plenty of other creatures to eat them up, and in this instance the predaceous animalcule, already described, was not the only enemy the little green globes had to suffer from, as two sorts of rotifer were frequently met with. One of these was a very handsome and singular creature, which in some positions had the general contour of a cockatoo, only that the legs were wanting, and the head exhibited a monkey face. The "wheels" were represented by ciliary tufts, and two bright red eyes twinkled with a knowing look. From each shoulder proceeded a long curved spine, and about two thirds down the body, and lying between the two long spines, a shorter one was articulated, which followed the same curve. A gizzard was busy in the breast, and the body terminated in two short toes, which grasped a large round egg. Whenever the cilia were drawn in, the three spines were thrown up; but they had an independent motion of their own, and every now and then were jerked suddenly and violently back, which occasioned a rapid change in the creature's position. The gizzard appeared to consist of two rounded masses, having several ridges of teeth, which worked against each other something like the prominences of a coffee-mill. From the three spines, this animal was a *Triarthra*, or Three-limbed Rotifer, but the position of the spines, and the toes, made it differ from any species described in the 'Micrographic Dictionary,' or in Pritchard.

Whether or not this species is to be regarded as having a lorica or not, must depend upon the precise meaning attached to that word. At any rate the integument was much firmer than in many of the rotifers, and gave an efficient support to the spines which a mere skin could not do. As Mr. Gosse remarks of an allied genus, the *Polyarthra*, or Many-limbed Rotifer, this creature could not be investigated without coming to the conclusion "Here again we have true jointed limbs;" a fact of great importance in determining the zoological rank of the family, and in supporting Mr. Gosse's view some at least bore a strong affinity with the group of *Arthropoda*, of which the insects are the principal representatives.

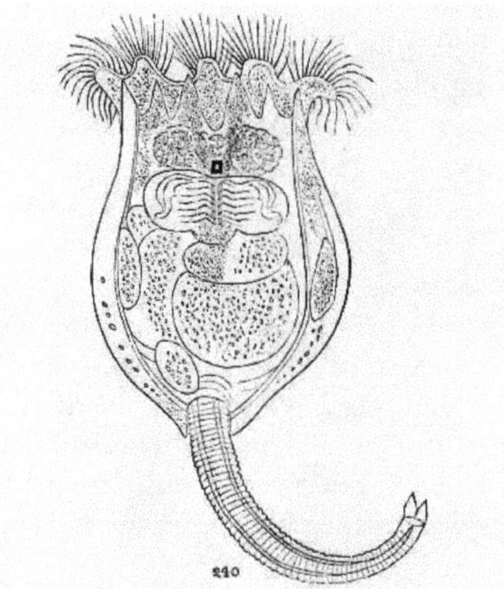

Brachionus urceolaris. This drawing has been accidentally reversed by the engraver, which alters the relative place of the internal organs.

Another rotifer of even greater interest, which was busy among the Cryptomonads, was the Brachion, or "Pitcher Rotifer" (Brachionus). The members of this genus will frequently reward the searcher into pond-life. Their main characteristic is a cup or pitcher-shaped lorica, which is cut or notched at the top into several horns or projections, the number of which indicates the species; while two or more similar projections ornament the bottom. This lorica is like the shell of a tortoise open at both ends; from the top an extremely beautiful wreath of cilia is protruded, and also some longer and stiff cilia, or slender spines, which do not exhibit the rotatory movement. The ciliary apparatus is in reality continuous, but it more often presents the appearance of several divisions, and the lateral cilia frequently hang over the sides. From the large size of each cilium they are very favorable creatures for exhibiting the real nature of the action, which gives rise to the rotatory appearance, and which can be easier studied than described. By movements, partly from their base, and partly arising from the flexibility of their structure, the cilia come alternately in and out of view, and when set in a circular pattern, the effect is amazingly like the spinning round of a wheel. The internal arrangements of the Brachiones are finely displayed, and they have a most aldermanic allowance of gizzard, which extends more than half way across each side of the median line, and shows all the portions described by Mr. Gosse. As the joints of this machine move, and the teeth are brought together, one could fancy a sound of mill-work was heard, and the observer is fully impressed with a sense of mechanical power.

When the creature is obliging enough to present a full front view, her domestic economy is excellently displayed. The prey that is caught in her whirlpool is carried down by a strong ciliary current to the gizzard, which may be often seen grappling with objects that appear much too big for its grasp; and Mr. Gosse was lucky in witnessing an attempt to chew up a morsel that did actually prove too large and too tough, and which, after many ineffectual efforts, was suddenly cast out. As soon as food has passed the gizzard, it is assisted in its journey by more ciliary currents, which are noticeable in the capacious stomach, in the neighbourhood of which the secreting and other vessels are readily observed. Just over the gizzard blazes a great red eye, of a square or oblong form, and it reposes upon a large mass of soft granular-looking brain, which well justifies Mr. Gosse's epithet "enormous." Whether this brain is highly organized enough to be a *thinking* apparatus, we do not know, but it is evidently the cause of a very vigorous and consentaneous action of the various organs the Brachion possesses.

A description of the Brachion would be very incomplete if it omitted that important organ the tail, which in this family reaches the highest point of development. It is a powerful muscular organ, of great size in proportion to the animal, capable of complete retraction within the carapace, and of being everted wholly, or partially, at will. It terminates in two short conical toes, protruded from a tube-like sheath, and capable of adhering firmly even to a substance so slippery as glass. This tail may be observed to indicate a variety of emotions, if we can ascribe such feelings to a rotifer, and it answers many purposes. Now we see it cautiously thrust forth, and turned this way and that, exploring like an elephant's trunk, and almost as flexible. Now it seizes firm hold of some substance, and anchors its proprietor hard and fast. A few moments afterwards it lashes out right and left with fury, like the tail of a cat in a passion. Then again it will be retracted, and a casual observer might not imagine the Brachion to be furnished with such a terminal implement.

The Brachiones may often be seen with one or more large eggs stuck about the upper part of the tail, and others may be discerned inside. One specimen before us has three eggs attached to her in this way. They are large oval bodies, with a firm shell. These creatures differ very much in appearance, according to the direction in which they are seen, and a side view makes them look so different from a full front or back aspect, that it would be easy to suppose another animal was under observation. The extent to which the ciliary apparatus is protruded, and the pattern it forms likewise differs continually; and hence no drawing, however correct, is sure to resemble the arrangement that may be presented to the observer's eye. But however our little "Pitcher" may be viewed, it is sure to prove a spectacle of interest and delight.

CHAPTER IX.

SEPTEMBER.

Microscopic value of little pools—Curious facts in appearance and disappearance of Animalcules and Rotifers—Mode of preserving them in a glass jar—Fragments of Melicerta tube—Peculiar shape of Pellets—Amphileptus—Scaridium Longicaudum—A long-tailed Rotifer—Stephanoceros Eichornii—A splendid Rotifer—Its gelatinous bottle—Its crown of tentacles—Retreats on alarm—Illumination requisite to see its beauties—Its greediness—Richly-coloured Food—Nervous ganglia.

SCATTERED about Hampstead Heath are a number of little pools, not big enough to be dignified by the name of ponds. They are generally surrounded by furze bushes, and would escape attention if not actually looked for. Those which are mere puddles, and have only a brief existence in rainy weather, seldom reward the labour of investigation; but others are permanent, except after prolonged drought, and afford convenient situations for the growth of confervæ, star-weed, and other plants. These will nearly always repay the microscopic collector during the winter, when he must break the ice to get at their contents; in spring, when long chains of frog-spawn afford ocular evidence of the prolific properties of the Batrachian reptiles; and in summer, when they afford both shade and sunshine to their numerous inhabitants. Small beetles, water-spiders, larvæ of gnats, and other insects, rotifers, including the tubicolar sorts, and several varieties of infusoria may be expected and generally found. There is, however, a curious fact about ponds, big and little, which Pritchard remarks upon in his 'Infusoria,' and which corresponds with our own experience, that those which have proved to be well stocked with any particular creature during one year, will very likely contain none of it in the next. There are of course exceptions to this rule, but we have often been astonished and disappointed at finding the complete change, both in populousness and population, that a revolution of twelve months will make; and it would be extremely interesting to notice the changes that took place during a term of years.

Such researches might unfold some unexpected laws in the succession of infusorial life. Those germs which are most widely diffused, will be the most likely to be developed in any mass of convenient water; but how and why the rarer forms come and go is very imperfectly understood. Slight modifications in surrounding circumstances will materially affect the result. Thus, if we bring home a handful of conferva, and a few water-plants of higher organisation, such as duckweed and anacharis, and place the whole in a glass jar full of pond-water, we shall at first have a good stock of objects; but they will usually grow less and less, until scarcely anything is left. If, however, we introduce a few pieces of straw, or a tiny wisp of hay, we shall succeed much better, and not only preserve our population longer, but enjoy a succession of animated crops. Extensive decomposition of vegetable matter kills off all but certain families, such as Paramecia, who enjoy it; on the other hand, too little decomposition proves fatal to some creatures, by depriving them of their food, and when they have died off, those who depended upon them for a living, die too. Different vegetables in decomposition suit different creatures, and hay and straw in that state seem to please the largest number. An animalcule tank will succeed best when it contains two or three kinds of growing plants, which oxygenize the air, and a moderate variety of decomposing organisms will supply food without making the water offensive.

From these considerations it will be apparent that not only the nature of the vegetation of a pond, which is often changed by accidental circumstances, but also the quality of the odds and ends that the winds may blow into it, or which may fall through the air, will do much to determine the character and number of its inhabitants, while the quantity of shade or sunshine it enjoys, will also exercise an important influence. Hay and other infusions have from the beginning of microscopic investigations been employed to obtain the creatures which the Germans call "Infusions thierchen" (infusion animalcules), and the English "Infusoria;" but very little has yet been done in the way of their scientific culture and management.

To return from this digression to our little Hampstead ponds, we obtained from one, in September, that was full of star-weed, a number of sugar-loaf bodies, adhering to one another, and of a pale yellow brown colour. The specimens first examined looked complete in themselves, and were taken for eggs of some water creature. Further search, however, disclosed aggregations of similar sugar-loaves that had evidently formed part of a tubular structure, and the idea at once occurred that they were fragments of a Melicerta tube, a conclusion that was verified by finding some tubes entire and a dead Melicerta in the rubbish at the bottom. All the specimens of Melicerta tubes we had hitherto examined were composed of *rounded* pellets, but these were made of pointed cones or sugar-loaves, with the points projecting outwards from the general surface. In Pritchard's 'Infusoria,' these pellets are described "as small lenticular bodies." The 'Micrographic Dictionary' states that the tubes of the Melicerta are composed of "numerous rounded or discoidal bodies;" and Mr. Gosse, in his 'Tenby,' which contains an admirable description, and an exquisite drawing of this interesting rotifer, calls the pellets "round."

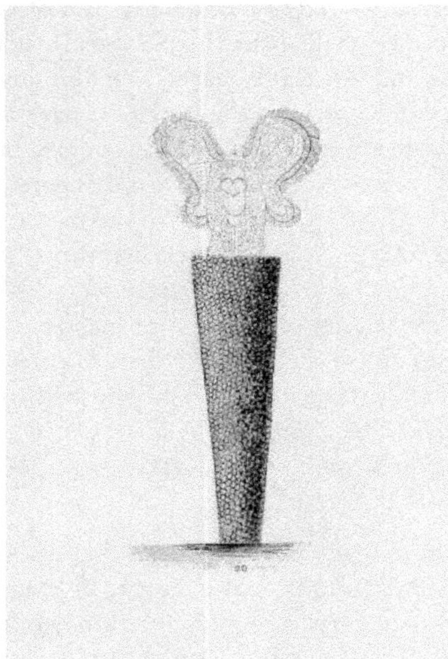

Melicerta ringens.

Not being able to obtain a living specimen of the Melicerta, who made her tube of long sugar-loaves, I could not tell whether she differed in structure from the usual pattern of her race, but the general appearance of the dead body was the same. It is possible that these creatures possess some power of modifying the form of their singular bricks, or they may at different ages vary the patterns, which matters some fortunate possessor of a colony of these animals may be able to verify.

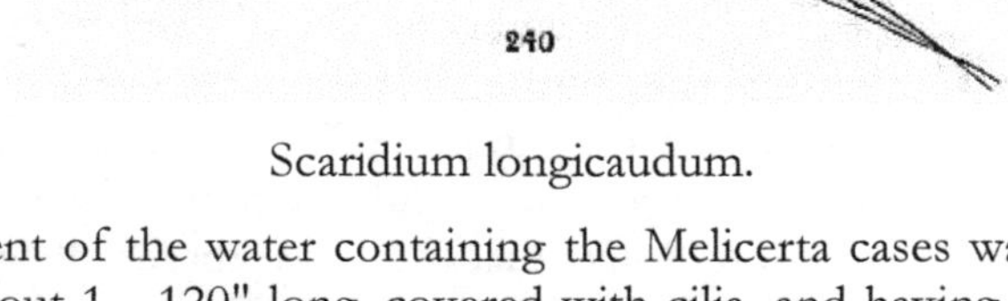

Scaridium longicaudum.

In the sediment of the water containing the Melicerta cases was found an animalcule about 1—120" long, covered with cilia, and having a proboscis seldom more than a quarter of the length assumed by the body, which continually changed its form, sometimes elongating, sometimes shortening, and often contracting one side into a deep fissure. It was, probably, an *Amphileptus*, though not precisely agreeing with any drawing or description I am acquainted with. Another inmate of the same water was a lively long-tailed rotifer, with a small oval body, a tuft of vibrating cilia and a curved bristle visible among them on one side. This creature had a jointed tail-foot, ending in two long style-shaped toes, and by means of this appendage executed rapid leaps or springs. It was the *Scaridium longicaudum*, and agreed in dimensions tolerably well with the size given in the books, namely, total length 1—72". With a power of five hundred diameters the muscles of the tail-foot presented a beautifully striated appearance.

Towards the end of the month I passed the Vale of Heath Pond, Hampstead, and although I had not gone out for the purpose of collecting, was fortunately provided with a two-dram bottle. Close by the path the *Anacharis alsinastrum* grew in profusion, quantities of water-snails crawled among its branches, and small fish darted in and out, threading their mazes with lightning rapidity. Thrusting a walking-stick among the mass of vegetation, a few little tufts were drawn up and carefully bottled, with the addition of a little water. Returning home, a few leaves were placed in the live-box, and on examination with the power of sixty diameters they disclosed a specimen of, perhaps, the most beautiful of all the rotifers, the *Stephanoceros Eichornii.* In this elegant creature an oval body, somewhat expanded at the top, is supported upon a tapering stalk, and stands in a gelatinous bottle, composed of irregular rings superimposed one upon the other, as if thrown off by successive efforts, the upper ones being inverted and attached to the body of the animal. But that which constitutes the glory of this little being is the crown of five tapering tentacles, each having two rows of long cilia arranged on opposing sides, but not in the same plane. The ordinary position of the tentacles is that of a graceful elliptical curve, first swelling outwards, then bending inwards, until their points closely approximate, but each is capable of independent motion, and they are seldom quiet for many minutes at a time. The cilia can be arranged in parallel rows or in tufts at the will of the creature, and their motion appears under control, and susceptible of greater modification than is exhibited by the ordinary infusoria.

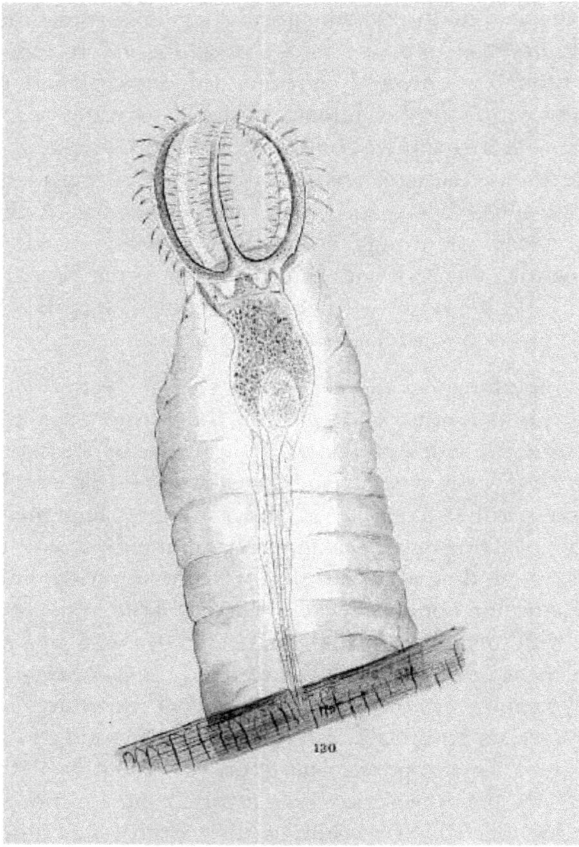

Stephanoceros Eichornii.

The Stephanoceros is a member of the Floscule family, but in all the specimens I obtained and watched for several weeks, there was an important difference in the relation of the tube to the creature. In the Floscules I had never seen anything like an adhesion between the tube and the animal, but in the Stephanoceros I noticed it continually, and always in the manner already described. Like the Floscule, the Stephanoceros is readily alarmed, and retreats into her house, carrying with her the invaginated portion. In the last edition of 'Pritchard's Infusoria,' this case is spoken of as apparently not tubular, but a solid gelatinous mass, enveloping the animal as high up as the base of the rotatory arms. It is very likely that specimens at different ages, and possibly in different seasons, may vary in the structure of their abodes; but I am not able to concur in the preceding account, as all the tubes I examined resembled sacks turned in at the mouths, and attached to the shoulders only of their inmates; and on one occasion I was able to look down into a deserted tube, which had not collapsed, as it would have done if it had been merely a solid gelatinous mass.

Like the Floscule, the Stephanoceros only reveals her beauties under careful illumination. A direct light renders them invisible, and only when the requisite obliquity has been obtained, does the exquisite character of the structure become displayed. The dark-ground illumination is very useful, and makes the ciliary action very distinct. At times a view can be obtained, in which the cilia of perhaps a single tentacle are all ranged like the steel springs of a musical box. For a moment they are quiescent, and then they vibrate in succession, each moving thread sparkling in the light. With a clumsy mode of lighting them, the cilia look like stumpy bristles, and are often so drawn; but precisely the right quantity of light coming in the right direction, makes them appear more numerous, and much longer than would at first be supposed. When well exhibited the tentacles have a lustre between glass and pearl; the body, in a favorable specimen, is like a crystal cup, and the food, usually composed of small red and green globes, glows like emeralds and rubies, as if in the height of luxury the little epicure had more than rivalled Cleopatra's draught, and instead of dissolving, swallowed its jewelry whole. So lustrous and varied in colour is the whole appearance of the animal under these circumstances, that it is frequently alluded to by one of our first artists, to whom it was displayed.

It is said by some authors that the tentacles are used to seize prey. This never occurred under my observation, although their basal portions are often approximated when an object is forced down to the grinding apparatus below. The Stephanoceros is a ravenous feeder, and swallows a variety of creatures. Green vegetable monads, rich red and brown globes of similar characters, and any animalcule that comes in her way is acceptable; and even good-sized rotifers do not escape her all-consuming maw. On one occasion I noticed one of the loricated sort, more than half as long as one of her tentacles, rapidly swallowed, and passed downwards without attempting to escape. Objects much too big for the gizzard are often gulped down, and probably receive a preliminary softening and maceration in the crop. Very often, when food is plentiful, the creature is filled to the brim, but still endeavours to continue her abundant meal. From the presence of large quantities of food and the density of the integuments, the gizzard cannot always be seen; but in favorable specimens its teeth may be observed busily at work.

At the base of the tentacles small masses of matter may be discovered, which are probably nervous ganglia, and other organs; and Ehrenberg discovered small vibrating bodies, supposed to be connected with the function of respiration. A single egg, as shown in the annexed drawing, is often found, and the ovarian is said to develop but few at a time. Two red eyes are found in young specimens, but in adults they either disappear or are not conspicuous. The Stephanoceri are sociable animals, and when one is found, others are probably near at hand. Several may often be discovered on the same branch of a small water-plant, of various dimensions, and in different stages of growth. The full size is about 1—36" in height, and from its magnitude care is required not to crush it in the live-box. When specimens are plentiful, some should be placed in that convenient receptacle; and others with the plant on which they are growing, in a glass cell or trough, where they have more room to display their motions, and can with fresh supplies of water, be preserved for days and weeks. With occasional renewals from one pond, I was able to keep up a stock for about three months, and never had objects which gave more pleasure to myself or to my friends.

CHAPTER X.

OCTOBER.

Stentors and Stephanoceri—Description of Stentors—Mode of viewing them—Their abundance—Social habits—Solitary Stentors living in Gelatinous caves—Propagation by divers modes—Cephalosiphon Limnias—A group of Vaginicolæ—Changes of shape—A bubble-blowing Vorticella.

OCTOBER, the finest of our autumn months, is noted for usually granting the inhabitants of our dripping climate about twenty pleasant sunshiny days, and it is probably on this account somewhat of a favourite with the infusorial world, although the cold of its nights and early mornings thins their numbers, which reach a maximum in the summer heat. Even in the dismal year 1860, October maintained its character, and afforded a great many opportunities of animalcule hunting, during which a constant supply of Stephanoceri were readily obtained, together with swarms of *Stentors*, which are not exceeded in interest by any of the Ciliated Protozoa. The Stentors were abundant on the same weed (*Anacharis*), that formed the residence of the Stephanoceri, and might be seen in large numbers hanging from it like green trumpets, visible to the unassisted eye. In the 'Micrographic Dictionary' they are said to belong to the Vorticella family, which has already given us several beautiful objects, and possess a marvellous power of changing their shape. It is, however, better to follow Stein, who separates them from the Vorticellids and ranges them in his order Heterotricha, as they have two distinct sets of cilia, small ones covering the body and the larger ones round the mouth. Those before us are named after this property *Stentor polymorphus*,[18] or Many-shaped Stentors, and owe their exquisite tint to numberless green vesicles, or small cavities filled with colouring matter like that of plants. This, however, is not essential to the species which may often be found of other hues. In size this Stentor varies from a hundred and twentieth to one twenty-fourth of an inch. It is entirely covered with fine cilia, disposed in longitudinal rows, and round the head is a spiral wreath of larger and very conspicuous cilia leading to the mouth.

[18] See Frontispiece.

Having observed the abundance of these creatures, a few small branches to which they were appended, were placed in the glass trough, and viewed with powers of sixty and one hundred linear. Some had tumbled down as shapeless lumps, others presented broad funnel-shaped bodies; while others stretched themselves to great length like the long, narrow post-horns which still wake the echoes of a few old-fashioned towns. The ciliary motion of the elegant wreath was active and rapid, causing quite a stir among all the little particles, alive and dead; and when the right sort of food came near the corkscrew entrance to the mouth, down it went, and if conspicuous for colour, was subsequently seen apparently embedded in little cavities, which Ehrenberg supposed were separate stomachs, although that theory is now rejected. One advantage of viewing these objects in a sufficient quantity of water, to leave them in freedom, is that they frequently turn themselves, so that you can see right down into them; and the drawing given in the frontispiece represents such a view, which is the most favorable for the exhibition of the mouth. To make out the details of their structure, to see the nucleus and other organs, the flattening in the live-box is useful, and it enables much higher powers to be employed.

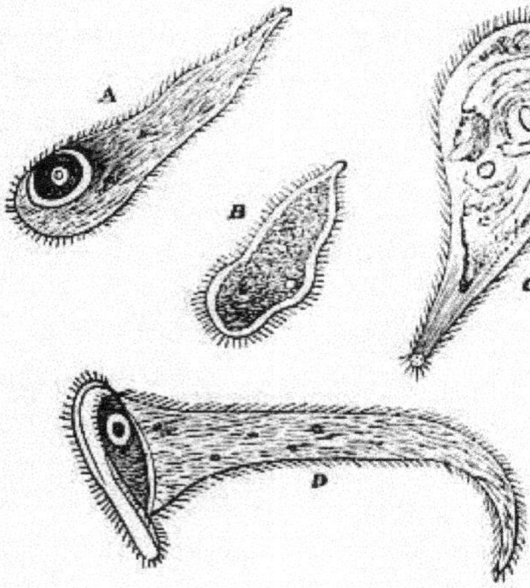

A, B, C, D, Stentor polymorphus in different degrees of expansion. A large specimen is one twenty-fourth of an inch long.

After leaving the Anacharis in a glass jar for a few days, the Stentors multiplied exceedingly; some clung to the sides of the vessel in sociable communities, others hung from the surface of the water, and crowds settled upon the stems, visibly changing their tint, as the Stentor green was much bluer than that of the plant. Scores swam about in all sorts of forms. Now they looked like cylindrical vessels with expanding brims, now globular, now oddly distorted, until all semblance of the original shape was lost. Many were found in shiny tubes, but these were never so lively or green as the free swimmers, but mostly of a dingy dirty hue.

These housekeepers were more timid and cautious than the roving tribe. They came slowly out of their dens, drew back at the slightest alarm, never took their tails from home, and only extended their full length when certain not to be disturbed. Some authors have thought they only take to private lodgings when they feel a little bit poorly, but others dispute this opinion, and I do not think it is correct. I have found these Stentors at all seasons, from January to the autumn, but they are never so numerous, nor aggregated in numbers like the roving sort. Whether they are old folks, who are tired of the world and its gaieties, and devote the remainder of their lives to contemplation, or whether they are bachelors disappointed in love, I am unable to say; but they are very inferior in beauty to the "gay and glittering crowd."[19]

[19] Stein says the colourless variety of S. Polymorphus is sometimes found with a tube, and the S. Rössellii very frequently so provided.

For some weeks my Stentors abounded, and then most of them suddenly disappeared. They could not have "moved," but probably "went to smash" by a process peculiar to infusoria, and which Dujardin politely describes as "diffluence." This mode of making an exit from the stage of life is more tragical than the ripping up so fashionable in Japan. The integument bursts, and its contents disperse in minute particles, that in their turn disappear, and scarcely leave a "wrack behind."

The Stentors obey the injunction to "increase and multiply" by self-division, which Stein says is always oblique, and the nucleus, which plays such an important part in infusoria, is band-like, moniliform (bead-shape), or round. When an animalcule increases by self-division, a portion of the nucleus goes with each part, and it is probably the organ which stimulates the change. It is also concerned in other modes of propagation. "The anus is situated on the back close beneath the ciliary circle;" and the "contractile vesicle on a level with the ciliary wreath." Stein records that in November, 1858, he met green Stentors (*Polymorphus*) encysted, and he figures one in a gelatinous flask having a stopper in its narrow neck.

Before closing our account of the Stentor, let us revert a moment to the ciliary wreath, as it may be made the subject of a curious experiment. If, for example, the cilia are viewed at right-angles to their length, they will seem to form a delicate frill, in which a quivering motion is perceived. But if the table is shaken by a sharp blow, the frill is thrown into waves, or takes the form which washerwomen give to certain female articles by the use of the Italian iron, and the ciliary motion is thus made to take place in different planes, and rendered strikingly apparent.

One day turning over the Anacharis in search of subjects, a small brown tube was noticed, from which a glassy rod protruded like the feeler of a rotifer. Keeping the table quiet, and watching the result, was soon rewarded by a further protrusion of the feeler, accompanied by a portion of the body of the inmate of the tube. The feeler was thrust on this side and on that, as if collecting information for its proprietor, who, I suppose, was satisfied with the intelligence, and gradually extended herself, until she stood out two thirds in length beyond the tube, and set two lobes of one nearly continuous ciliary organ in rapid motion. Sometimes the creature, *Cephalosiphon limnias*, bent its neck, if I may so speak, to the right, and sometimes to the left, and sometimes stood upright, when the true form of the ciliary apparatus could be seen. The tube of this creature was opaque, from the adhesion of foreign matter, and presented an untidy appearance, strangely contrasting with the clear, neat bottles of the Floscules. These Cephalosiphons are very whimsical in their ways, and many that were sent to different observers never exhibited their ciliary wreaths, but performed sundry antics, disguising their true shape.

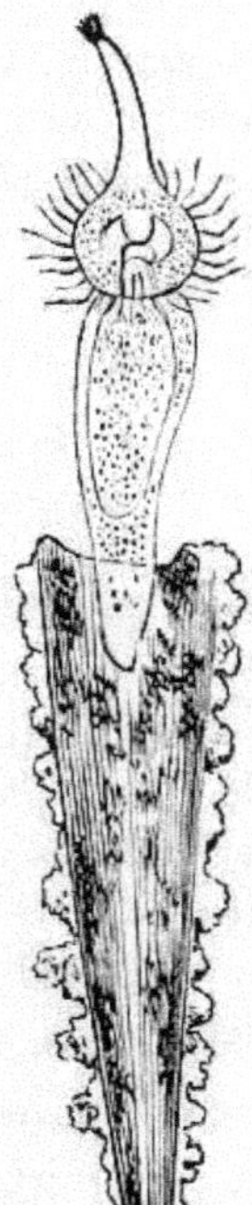

Cephalosiphon limnias.

Somewhat like the Cephalosiphon, though much commoner and without the siphon, is Limnias ceratophylli, which every collector is sure to meet. The length of the Limnias varies, according to Pritchard, from 1—20" to 1—40". Our Cephalosiphon, when fully extended and magnified one hundred and eighty linear, looked about three inches and a half long, and was therefore very small. Just below the ciliary lobes the gizzard was seen, with its toothed hammers working one against the other. The masticatory organ differs from the typical form, as represented in the Brachion; and Mr. Gosse observes of Limnias that "each *uncus* forms, with its *ramus*, a well-defined mass of muscle enclosing the solid parts, and in form approaching the quadrature of a globe. Across the upper surface of the mass the *uncus* is stretched like three long parallel fingers, arched in their common direction, and imbedded in the muscular substances, their points just reaching the opposing face of the *ramus*, and meeting the points of the opposite *uncus* when closed."[20]

[20] The terms *uncus*, *ramus*, etc., have been explained in Chapter II, page 28.

There is no connection between Limnias or Cephalosiphon and their tubes, except that of simple adhesion, which takes place by means of the end of their foot-stalks.

In a former chapter we have described an interesting relation of the Vorticella, the Cothurnia, whose elegant crystal vases form a very artistic abode, characterised by possessing a distinct foot. Other species of the same family inhabit vases which have no foot or stalk, or live in gelatinous sheaths less accurately fashioned. Sometimes these creatures are obliging enough to conform to the specific descriptions which eminent naturalists have given of them, and also to the characters which the authorities have assigned to the different genera in which they have been grouped, but the microscopist will often meet with difficulties in the way of classification.

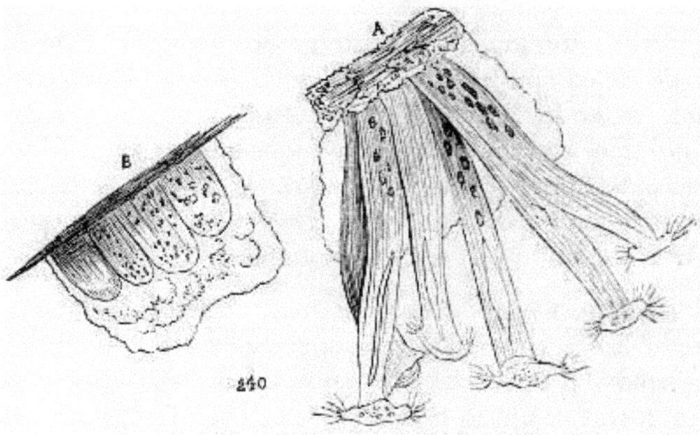

Vaginicola (?) (A, elongated; B, retracted.)

Attached to a piece of weed were a number of cylindrical masses of brownish jelly, with rounded tops, and situated in an irregular and very transparent sheath, about twice as high as themselves. Presently they all rose up to four times their previous height, put forth a beautiful crown of vibrating cilia, and opened a sort of trap-door to their internal arrangements. In this position they had a long cylindrical form, gracefully curved, but of nearly equal width from the mouth to the base, and they readily imbibed particles of carmine, which tinged sundry little cavities with its characteristic hue. The slightest disturbance caused the ciliary wreaths to be drawn in, and the bodies to be retracted, and descend into their house like a conjuring toy, until the appearance first described was reproduced.

The general form and structure of these objects was like the drawings usually given of *Vaginicola*, which is said not to exist in groups, although two individuals are commonly found in one well-shaped cell. These creatures, however, did not taper towards the base as Vaginicolæ generally do, and perhaps they became aware of this defect in their figures, for after a day or two a change appeared, and they assumed a more graceful form by swelling out in the middle, and then growing slender down to the bottom, very much like the pattern given by glass-blowers to little vases of flowers.

It is very important to note the changing appearance of animalcules, and where the same individuals can be observed from day to day, these will often be found considerable. It is probable that when such particulars are fully known, the number of species will be greatly reduced, and the study of these organisms considerably simplified. I have called the animals just described *Vaginicolæ*, but the reader must be prepared to find similar bodies, inhabiting well-formed vases, either solitarily or in couples, the latter condition arising from the fission of one individual without a corresponding division of the abode.

For a few weeks I continually met with groups living as I have described, in what may be called amorphous cells, which were often so nearly like the surrounding water in refracting power, as to be discerned with some difficulty. No trace could be seen of divisions into separate cells, but they all appeared to live happily together in one room, and if one went up all went up, and if one went down all went down, as if their proceedings were regulated by a community of sensation or will.

Another little curiosity was a transparent cup upon a slender stem, which stood upright like a wineglass, and supported on its mouth a transparent globe. By removing a leaf which prevented the stalk being traced to its termination, it was found to be a Vorticella, and after two hours the globe was partially drawn in, and reduced in size. Why the creature was engaged in blowing this bubble I do not know, and have not met with another instance of such conduct.

CHAPTER XI.

NOVEMBER.

Characteristics of the Polyzoa—Details of structure according to Allman—Plumatella repens—Its great beauty under proper illumination—Its tentacles and their cilia—The mouth and its guard or epistome—Intestinal tube—How it swallowed a Rotifer, and what happened—Curiosities of digestion—Are the tentacles capable of Stinging?—Resting Eggs, or "Statoblasts"—Tube of Plumatella—Its muscular Fibres—Physiological importance of their structure.

DURING the fag end of last month I observed some fragments of a new creature among some bits of Anacharis, from the Vale of Heath Pond, and searched for complete and intelligible specimens without effect. Luckily one evening a scientific neighbour, to whom I had given some of the plant for the sake of the beautiful *Stephanoceri* which inhabited it, came in with a glass trough containing a little branch, to which adhered a dirty parchment-like ramifying tube, dotted here and there with brown oval masses, and having sundry open extremities, from which some polyp-shaped animals put forth long pearly tentacles margined with vibrating cilia, and making a lively current. The creatures presented an organization higher than that of polyps, for there was an evident *differentiation* and complication of parts. They belonged to the *Polyzoa* or *Bryozoa*,[21] a very important division of the *mollusca*. The *Polyzoa* are chiefly marine, and the common "sea-mat," often erroneously treated as a *sea-weed*, is a well-known form. A species of another order often picked up on our coasts is the *Sertularia*, or Sea-Fir, composed of delicate branching stems of a horny-looking substance, which, under a pocket-lens, is found to contain an immense number of small cells inhabited by Polyps. It is instructive to compare the two and note how much more advanced in structure is the Polyzoon than the polyp.

[21] *Polyzoa* means "many animals," in allusion to their habit of living in association. *Bryozoa*, "moss-animals," from some forming cells having that appearance.

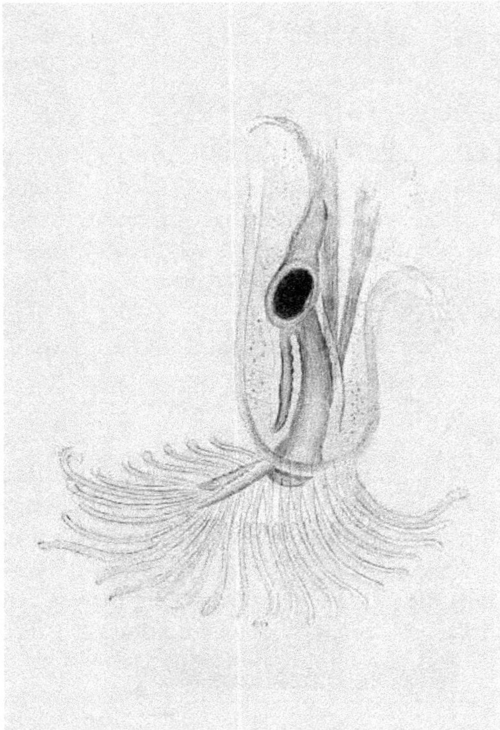

Plumatella repens. Single Polypide enlarged

Polyzoa were formerly associated with the polyps, to which they bear a strong superficial resemblance; but they are of a much higher degree of organization, as will be seen by comparing what has been said in a former chapter on the *Hydra*, with the description which we now proceed to abridge from Dr. Allman's splendid monograph on the fresh-water kinds. In order to get a general conception of a Polyzoon, the Professor tells us to imagine an alimentary canal, consisting of œsophagus, stomach, and intestine, to be furnished at its origin with long ciliated tentacles, and to have a single nervous ganglion on one side of the œsophagus. We must then conceive the intestine bent back till its anal orifice comes near the mouth; and this curved digestive tube to be suspended in a bag containing fluid, and having two openings, one for the mouth and the other for the vent. A system of muscles enables the alimentary tube to be retracted or protruded, the former process pulling the bag in, and the latter letting it out. The mouth of the bag is, so to speak, tied round the creature's neck just below the tentacles, which are the only portions of it that are left free. The investing sack has in nearly every case the power of secreting an external sheath, more or less solid, and which branches forming numerous cells, in which the members of the family live in a socialistic community, having, as it were, two lives, one individual, and the other shared in common with the rest.

The whole group of tubes and cells, whatever may be the form in which they are aggregated, is called the *Polypary*, or, as Dr. Allman prefers, the *Cœnœcium* (common house); the creature he names a *Polypide*[22] (polyp-like); and the disk which bears the tentacles *Lophophore* (crest-bearer). There are some more hard words to be learnt before the student can enjoy himself scientifically among the Polyzoa, and we shall be compelled to employ some of them before we have done; but will now endeavour to describe what was presented to our view by the specimen obtained from the Hampstead Pond.

[22] *Polyzoon* is preferable, as avoiding confusion with *polypite*, used for another class of object.

The general aspect of a branch of *Plumatella repens*—the creature we have to describe—is given in the drawing annexed. When all was quiet, the mouths of the bags belonging to each cell were slowly everted, and out came a numerous bundle of tentacles, which were either spread like the corolla of a flower, or permitted to hang dishevelled like the snake-locks of Medusa. We will suppose these organs symmetrically expanded, and that we are looking down upon them with a magnifying power of sixty diameters, the light having been carefully adjusted by turning the reflecting mirror a little on one side, to avoid a direct glare. The tentacles, each of which curves with a living grace, and displays an opaline tint in its glassy structure, do not form a complete circle, for at one place we discern two slightly diverging arms of the disk, or frame (Lophophore) from which they grow.

These arms support tentacles on each side, and leave a gap between, so that the whole pattern is *crescentic*, or crescent-shaped, and not circular. Extending as far as the points of the arms, and carried all round the crescent, is an extremely delicate membrane, like the finest gauze, which unites all the tentacles by their basal portions, and makes an elegant retreating curve between every two. Each tentacle exhibits two rows of cilia, which scintillate as their vibrations cause them to catch the light. The motion of the cilia is invariably *down* one side and *up* the other, the current or pattern being carried on from one tentacle to the other, all through the series. This characteristic, and the facility with which each cilium can be distinguished, gives great interest and beauty to the spectacle of this wonderful apparatus, by which water-currents are made to bathe the tentacles, and assist respiration, and also to carry food towards the mouth, over which a sort of finger or tongue is stretched to guard the way, and exercise some choice as to what particles shall be permitted to pass on. This organ is called the *epistome*, from two Greek words, signifying "upon the mouth."

If the cell is an old one, it may be covered with so much extraneous matter as to obscure the economy within; but we are fortunate in having a transparent specimen before us, through which we can see all that goes on. The alimentary tube, after forming a capacious cavity, much longer than it is broad, turns round and terminates in an orifice near the mouth, and just below the integuments. When refuse has to be discharged, this orifice is protruded; and after the operation is over, it draws back as before. Long muscles, composed of separate threads or fibres, pull the creature in and out of its cell, and at the part where the stomach ends, and the intestine turns round, is attached a long flexible rope, called the *funiculus*, which goes to the bottom of the cell. The passage of the food down to the stomach, its digestion, and the eviction of the residue, can all be watched; and when a large morsel is swallowed, the spectacle is curious in the extreme.

One day a polyzoon caught a large rotifer, (R. *vulgaris*,) which, with several others of its tribe, had been walking over the *cœnœcium*, and swimming amongst the tentacles, as if unconscious of danger. All of a sudden it went down the whirlpool leading to the mouth, was rolled up by a process that could not be traced, and without an instant's loss of time, was seen shooting down in rapid descent to the gulf below, where it looked a potato-shaped mass, utterly destitute of its characteristic living form. Having been made into a bolus, the unhappy rotifer, who never gave the faintest sign of vitality, was tossed up and down from the top to the bottom of the stomach, just as a billiard-ball might be thrown from the top to the bottom of a stocking. This process went on for hours, the ball gradually diminishing in size, until at last it was lost in the general brown mass with which the stomach was filled. The bottom of the stomach seems well supplied with muscular fibres, to cause the constrictions by which this work is chiefly performed, and by keeping a colony for a month or two, I had many opportunities of seeing my Polyzoa at their meals.

When alarmed the tentacles were quickly retracted, but although these creatures are said to dislike the light, and usually keep away from it in their native haunts, my specimens had no objection to come out in a strong illumination, and seemed perfectly at their ease. They were indeed most amiable creatures, and never failed to display their charms to admiring visitors, who rewarded them with unmeasured praise. Twice I had an opportunity of observing an action I cannot explain, except by supposing either that the tentacles of the *Plumatella* have some poisonous action, or that rotifers are susceptible of fear. On these occasions the common rotifer was the subject of the experiment. First one and then another got among the tentacles, and on escaping seemed very poorly. One fellow was, to borrow a phrase from Professor Thomas Sayers, "completely doubled up," and two or three seconds—long periods in a rotifer's life—elapsed before he came to himself again.

By keeping a colony of the Plumatella for a few weeks in a glass trough, and occasionally supplying them with fresh water from an aquarium, containing the animalcules, they are easily preserved in good health, and as they develop fresh cells, the process of growth may be readily watched. This production of fresh individuals enlarges the parent colony, but could not be the means of founding a new one, which is accomplished by two other modes. A little way down the cells Professor Allman discovered an ovary attached to the internal tube by a short *peduncle*, or foot stalk, while a testis or male generative organ is attached to the *funiculus*, or "little rope," we have already described.

July and August are the best times for observing the ovaries, and they are most conspicuous in the genera *Alcyonella* and *Paludicella*. True eggs are developed in the ovaries in a manner resembling this mode of multiplication in other animals; but there is another kind of egg, or, perhaps to speak more properly, a variety of bud, which is extremely curious. In looking at our specimens we noticed brown oval bodies in the cells; these, on careful examination, presented the appearance of the sketch. The centre is dark, covered with a network, which is more conspicuous in the lighter coloured and more transparent margins. These curious bodies are produced from the funiculus, and act as reserves of propagative force, as they are not hatched or developed until they get out and find themselves exposed to appropriate circumstances. Professor Allman names them *Statoblasts*, or stationary germs, and they bear some resemblance to what are called the "winter eggs" of some other creatures. The Professor was never able to discover any mode by which they were permitted to escape from the cells, and in our colonies none were allowed to leave their homes until the death of their parent, and the decomposition of its cell had taken place; a process which went on contemporaneously with the growth of new cells, until the plant on which the *cœnœcium* was situated, rotted away, and then unfortunately the whole concern went to pieces.

Plumatella repens on a leaf.

The tubes of the *Plumatella*, and of most other Polyzoa, are composed of two coats, called respectively *endocyst* and *ectocyst*, that is, "inner case" and "outer case." The first is vitally endowed, and exhibits vessels and muscular fibres. The second or outer case is thrown off by the first. It is a parchment-like substance, strengthened by the adhesion of dirt particles, and does not appear to exercise any vital functions, but to be merely a covering for protection. The inner layer terminates in the neck of the bag before described, as exserted when the polypide comes out, and inverted when it goes in. This mode of making a case or sheath by inversion of a bag is technically called *invagination*, and is readily seen in new and transparent cells.

The movement of *eversion*, or coming out, is chiefly produced by the contraction of the endocyst; while the *inversion*, or getting in again, is performed by the long muscles, which, when the animal is extended, are seen attached to it like ropes. Upon these muscles Professor Allman remarks that they are "especially interesting in a physiological point of view, as they seem to present us with an example of true muscular tissue, reduced to its simplest and essential form. A muscle may here be viewed as a beautiful dissection far surpassing the most refined preparation of the dissecting needle, for it is composed of a bundle of elementary fibres, totally separate from one another through their entire course." He further adds, "The fibres of the great retractor muscle are distinctly marked by transverse striæ;—a condition, however, which is not at all times equally perceptible, and some of our best observers have denied to the Polyzoon the existence of striated fibre."

We can confirm the fact of this sort of fibre being present, but we fancy a reader not versed in the mysteries of physiology exclaiming, 'What does it matter whether his fibres are striped or not?'

Physiologists used to suppose there was a strong and marked distinction and separation between *striped* muscles, that is, muscles the fibres of which exhibit transverse stripes when magnified, and those which do not. Kölliker, however, says this decided separation can no longer be maintained,[23] and he gives instances in proof of the connections that can be traced between the two forms. In the higher animals the striped muscles are the special instruments of *will*, and of movements that follow, or are accompanied by, distinct sensations. Striped fibre must be regarded as the highest form; and as a muscle of this sort contracts in length it increases uniformly in breadth.

[23] 'Manual of Human Microscopic Anatomy,' p. 63.

There are many other genera and species of fresh-water polyzoa besides the *Plumatella repens*, and they are found attached to sticks, stones, or leaves, generally to the under surface of the latter. They are all objects of great interest and beauty, which, whatever their diversity, conform sufficiently to one type that the student who has observed one, will easily recognise the zoological position of another. They should be viewed by transmitted and by dark-ground illumination, which produces very beautiful effects. To observe them in the performance of their functions, they require more room than the live-box can afford, but are well shown in the glass trough, whose moveable diaphragm enables them to be brought near enough to the object-glass, for the use of a power of about sixty linear for general purposes, and of from one to two hundred for the examination of particular parts. For a more detailed examination dissection must be employed, but all that we have mentioned can be seen without injury to the living animal, if specimens are kept till new cells are formed in water, which does not contain enough dirt to render their integuments opaque.

CHAPTER XII.

DECEMBER.

Microscopic Hunting in Winter—Water-bears, or Tardigrada—Their comical behaviour—Mode of viewing them—Singular gizzard—A compressorium—Achromatic condenser—Mouth of the Water-bear—Water-bears' exposure to heat—Soluble albumen—Physiological and chemical reasons why they are not killed by heating and drying—The Trachelius ovum—Mode of swimming—Method of viewing—By dark-ground illumination—Curious digestive tube with branches—Multiplication by division—Change of form immediately following this process—Subsequent appearances.

THERE is always satisfaction in finding a work accomplished; but the attempt to delineate some of the marvels of minute creation has been a pleasant one, and we approach the completion of our task of recording a *Microscopic Year* with something like regret. The dark, dirty December of the great metropolis may not seem a promising time for field excursions, but some ponds lie near enough to practicable roads and paths to render an occasional dip in them, not of ourselves, but of our bottles—an easy and not unpleasant performance; and if the weather is unusually bad, we can fall back upon our preserves in bottles and tanks, which seldom fail to afford something new, as we have been pretty sure to bring home some undeveloped germs with our stock of pond-water and plants, and even creatures of considerable size are very likely to have escaped detection in our first efforts at examination.

When objects are not over abundant, as is apt to be the case in the cold months, it is well to fill a large vial with some water out of the aquarium or other large vessel, and watch what living specks may be moving about therein. These are readily examined with a pocket-lens, and with a little dexterity any promising creature can be fished out with the dipping-tube. It is also advisable to shake a mass of vegetation in a white basin, as the larger infusoria, &c., may be thrown down; and indeed this method (as recommended by Pritchard) is always convenient. Even so small a quantity of water as is contained in a glass cell, appropriated to the continual examination of polyps or polyzoa, should be frequently hunted over with a low power, as in the course of days and weeks one race of small animals will disappear, and another take their place.

Following these various methods in December, we obtained many specimens; but the most interesting was found by taking up small branches of the Anacharis with a pair of forceps, and putting them into a glass trough to see what inhabitants they might possess. One of these trials was rewarded by the appearance of a little puppy-shaped animal very busy pawing about with eight imperfect legs, but not making much progress with all his efforts. It was evident that we had obtained one of the *Tardigrada* (slow-steppers), or Water-Bears, and a very comical amusing little fellow he was. The figure was like that of a new-born puppy, or "unlicked" bear cub; each of the eight legs were provided with four serviceable claws, there was no tail, and the blunt head was susceptible of considerable alteration of shape. He was grubbing about among some bits of decayed vegetation, and from the mass of green matter in his stomach, it was evident that he was not one of that painfully numerous class in England—the starving poor.

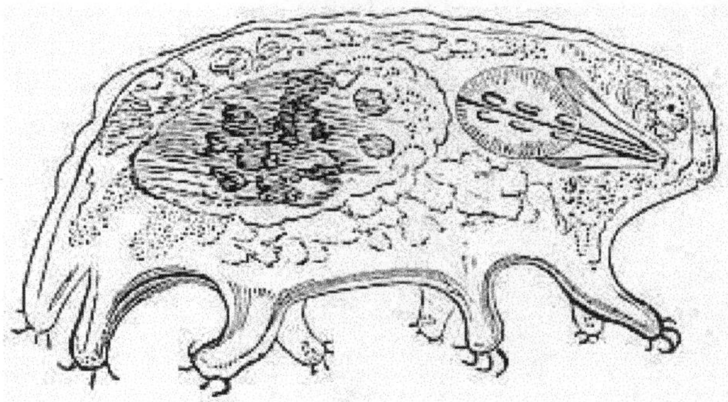

Water-Bear.

A power of one hundred and five linear, obtained with a two-thirds object-glass, and the second eye-piece, enabled all his motions and general structure to be exhibited, and showed that he possessed a sort of gizzard, whose details would require more magnification to bring out. Accordingly the dipping-tube was carefully held just over him, the finger removed, and luckily in went the little gentleman with the ascending current. He was cautiously transferred to a Compressorium,[24] an apparatus by which the approach of two thin plates of glass can be regulated by the action of a spring and a screw; and just enough pressure was employed to keep him from changing his place, although he was able to move his tiny limbs. Thus arranged, he was placed under a power of two hundred and forty linear, and illuminated by an achromatic condenser,[25] to make the fine structure of his gizzard as plain as possible. It was then seen that this curious organ contains several prominences or teeth, and is composed of muscular fibres, radiating in every direction. From the front of the gizzard proceed two rods, which meet in a point, and are supposed to represent the maxillæ or jaws of insects, while between them is a tube or channel, through which the food is passed. The mouth is *suctorial*, and the two horny rods, with their central piece or pieces, are protrusile. They were frequently brought as far as the outer lips (if we may so call the margins of the mouth), but we did not witness an actual protrusion, except when the lips accompanied them, and formed a small round pouting orifice. The skin of the animal was tough and somewhat loose, and wrinkled during the contractions its proprietor made. The interior of the body exhibited an immense multitude of globular particles of various sizes in constant motion, but not moving in any vessels, or performing a distinct circulation.

[24] The best forms of this instrument are made by Messrs. R. & J. Beck, the glass plates being held in their places by flat-headed screws, and not by cement. This plan was devised by the author, and makes it easy to renew the glasses when broken.

[25] The achromatic condenser is a frame capable of supporting an object-glass, lower than that employed for vision, through which the light passes to the object in quantities and directions determined by stops of various shapes. The appearances mentioned can be seen without it, though not so well.

My specimens had no visible eyes, and these organs are, according to Pritchard's book, "variable and fugacious." The same authority remarks, "In most vital phenomena they very closely accord with the rotatoria; thus like these they can be revived after being put into hot water at 113° to 118°, but are destroyed by immersion in boiling water. They may be gradually heated to 216°, 252°, and even 261°. It is also by their capability of resuscitation after being dried that they are able to sustain their vitality in such localities as the roofs of houses, where at one time they are subjected to great heat and excessive drought, and at another are immersed in water."

When vital processes are not stopped by excess of temperature, as is the case with the higher animals, the power of resisting heat without destruction depends upon the condition of the albumen. Soluble albumen, or, as it should be called, *Albuminate of Soda* (for a small quantity of that alkali is present and chemically united with it), after having been *thoroughly dried*, may be heated without loss of its solubility; although if the same temperature was applied before it was dry, that solubility would be destroyed, and it would no longer be a fit constituent of a living creature. As Dr. Carpenter observes, this fact is of much interest in explaining the tenacity of life in the Tardigrada.

The movements of the water-bears, although slow, evince a decided purpose and ability to make all parts work together for one common object; and as might be expected from this fact, and also from the repetition of distinct, although not articulated limbs, they are provided with a nervous apparatus of considerable development, in the shape of a chain of a ganglia and a brain, with connecting filaments. From these and other circumstances naturalists consider the Tardigrada to belong to the great family of *Spiders*, of which they are, physiologically speaking, *poor relations*. Siebold says "they form the transition from the Arachnoidæ to the Annelides."[26] Like the spiders they cast their skin; and, although I was not fortunate enough to witness this operation—called in the language of the learned *ecdysis*, which means putting its clothes off—I found an empty hide, which, making allowance for the comparative size of the creatures, looked tough and strong as that of a rhinoceros, and showed that the stripping process extended to the tips of the claws. The 'Micrographic Dictionary' states that the Tardigrada lay but few eggs at a time, and these are "usually deposited during the ecdysis, the exuviæ serving as a protection to them during the process of hatching." Thus Mrs. Water-Bear makes a nursery out of her old skin, a device as ingenious as unexpected. The water-bears are said to be hermaphrodites, but this is improbable.

[26] 'Anatomy of the Invertebrata,' Burnett's trans., p. 364.

The *Plumatella repens*, described in a former chapter, was kept in a glass trough, to which some fresh water was added every few days, taken from a glass jar that had been standing many weeks with growing anacharis in it. One day a singular creature made its appearance in the trough; when magnified sixty diameters it resembled an oval bladder, with a sort of proboscis attached to it. At one part it was longitudinally constricted, and evidently possessed some branched and complicated internal vessel. The surface was ciliated, and the neck or proboscis acted as a rudder, and enabled the creature to execute rapid turns. It swam up and down, and round about, sometimes rotating on its axis, at others keeping the same side uppermost, but did not exhibit the faintest sign of intelligence in its movements, except an occasional finger-like bend of the proboscis, upon which the cilia seemed thicker than upon the body. It was big enough to be observed as a moving white speck by the naked eye, when the vessel containing it was held to catch the light slantingly; but a power of one hundred and five was conveniently employed to enable its structure to be discerned. Under this power, when the animal was resting or moving slowly, a mouth was perceived on the left side of the proboscis, which was usually, though not always, curved to the right. The mouth was a round or oval orifice, and when illuminated by the parabola, its lips or margin looked thickened, and of a pale blue, and ciliated, while the rest of the body assumed a pinkish pearly tint.

Below the mouth came a funnel-shaped tube or œsophagus, having some folds or plaits on its sides, and terminating in a broad digestive tube, distinct from the nucleus, and ramifying like a tree. The constriction before mentioned, which was always seen in certain positions, although it varied *very considerably* in depth and width, drew up the integument towards the main trunk of the digestive tube, and thus the animal had a distinct ventral and dorsal side. The branches of the tube stopped somewhat abruptly just before reaching the surface, and were often observed to end in small round vacuoles or vesicles.

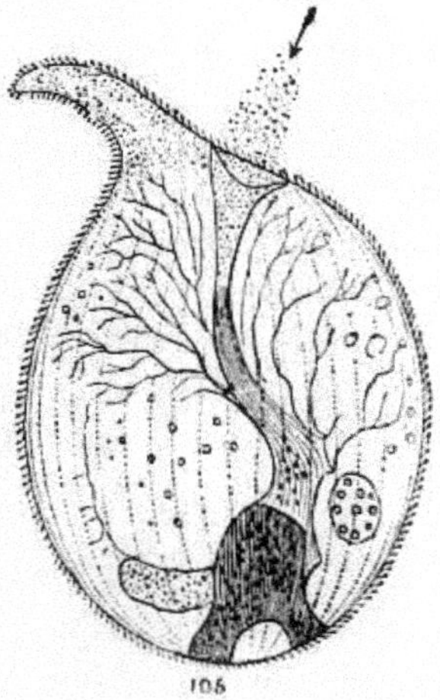

Trachelius ovum (slightly flattened).

At the bottom of the bladder, opposite the mouth, in some specimens were large round cavities or cells, filled with smaller cells, or partially transparent granules. These varied in number from one to two or three, and were replaced in other specimens by masses that did not present the same regular form or rounded outline. In one instance an amorphous structure of this kind gradually divided itself, and seemed in the course of forming two cells, but the end of the process was unfortunately not seen. The annexed drawing will readily enable the animal to be recognised. It shows the mouth very plainly, and a current of small particles moving towards it. The œsophagus terminates in a digestive tube, like the trunk of a tree, from which numerous branches spring. This arrangement is probably analogous to that of the phlebenterous mollusks described by Quatrefages, in which the ramifications of the stomach answer the purpose of arteries, and convey the nutrient fluid to various parts of the body. It is also likely that they minister to the function of respiration.

The cilia on the surface, which are arranged in parallel lines, are best observed when the animal is slightly flattened in a live-box; but this process produces a considerable derangement in the relative position of the internal parts, and they can only be well seen when it is immersed in plenty of water, and is polite enough to stand still, and submit his digestive economy to a steady gaze. The only way to succeed in this undertaking is to have a large stock of patience as well as a convenient cell or trough. The table must be kept steady, and the prisoner watched from time to time, and at last he will be found ready for display.

Pritchard says this animal, whose name is *Trachelius ovum*, is an inhabitant of stagnant bog water, and has been found encysted. My specimens could not be called plentiful, but for several weeks I could generally find two or three, by filling a four-ounce vial from the glass jar, and examining its contents with a pocket-lens. If none were present, another dip was made, and usually with success.

One evening I caught a good specimen by means of the dipping-tube, and cautiously let it out, accompanied by a drop of water, on the glass floor of the live-box. A glance with the pocket-lens showed all was right, and the cover was very gently put on, but it had scarcely touched the creature when it became crumpled up and in confusion. On one or two former occasions I had been unfortunate enough to give my captives a squeeze too much, with the usual result of a rupture of their integuments and an escape of globules and fluids from the regions within. Now, however, there was no such rupture and no such escape, but instead of a smooth, comely surface, my Trachelius had lost all title to his specific designation, *ovum*, for instead of bearing any resemblance to an egg, it was more like an Irishman's hat after having a bit of a "shindy" at Donnybrook Fair.

I was greatly puzzled with this aspect of things, and still more so when my deranged specimen twirled and bumped about with considerable velocity, and in all directions. Presently a decided constriction appeared about half-way below the mouth and proboscis, and in transverse direction. The ciliary motion became very violent in the lower half just below the constriction, while the proboscis worked hard to make its half go another way. For some minutes there was a tug of war, and at length away went proboscis with his portion, still much crumpled by the fight, and left the other bit to roam at will, gradually smooth his puckers, and assume the appearance of a respectable well-to-do animalcule.

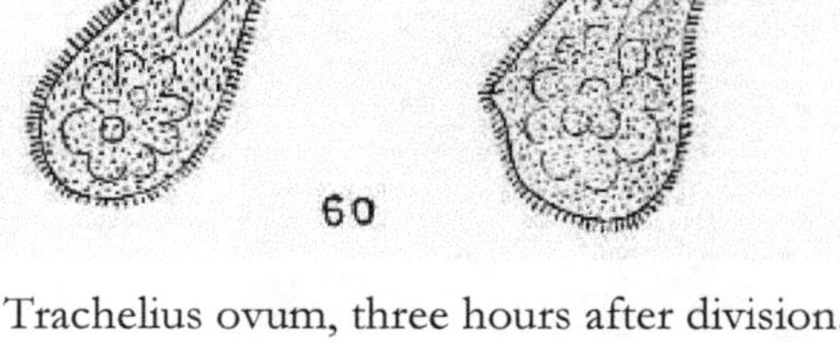

60

Trachelius ovum, three hours after division.

Three hours after the "fission" the proboscis half was not unlike the former self of the late "entire," but with diminished body and larger neck; while the remaining portion had assumed a flask form, and would not have been known by his dearest acquaintance. The portraits of the *dis-United States* were quickly taken, and, as bed-time had arrived, they were left to darkness and themselves. The next morning a change had come over the "spirit of their dream." Both were quiet, or sedately moving, and they were nearly alike. The proboscis fellow had increased and rounded his body, and diminished his nose; while Mr. Flask had grown round also, and evinced an intention of cultivating a proboscis himself. Twenty-seven hours after the separation, both had made considerable progress in arranging and developing their insides, which had been thrown into great confusion by the way in which the original animal had been wrenched in half, and in both a granular mass was forming opposite the mouth end. The proboscis portion, which may perhaps be termed the *mother*, was more advanced than her progeny, but both had a great deal to do if they meant to exhibit the original figure, and develop a set of bowels as elegantly branched. Whether they would have succeeded or not under happier circumstances I cannot tell, but unfortunately the Fate who carries the scissors cut short their days.

In all other animalcules in which I had observed the process of multiplication by self-division, it seemed to go on smoothly, and with no discomfort to either the dividend or the quotient, and it may be that in the fission of the *Trachelius ovum* I witnessed what the doctors would call a bad case. Indeed it may have been prematurely brought on, and aggravated by the squeeze in the live-box. It is, however, probable, from the stronger texture and greater organic development of this animalcule, that it does not divide so easily as the softer and simpler kinds.

Frequent examination of this animalcule has created a strong doubt in my mind whether it is rightly placed in our "systems." My own impression is that it belongs to a higher class.

CHAPTER XIII.

CONCLUSION.

THE creatures described in the preceding pages range from very simple to highly complicated forms, and in describing them some attention has been paid to the general principles of classification. The step is a wide one from the little masses of living jelly that constitute Amœbæ to the Rotifers, supplied with organs of sensation—eyes, feelers (calcars), and the long cilia in the Floscularians, which seem to convey impression like the whiskers of a cat—together with elaborate machinery for catching, grinding up, and digesting their prey, and which are also well furnished with respiratory and excretory apparatus, ovaries, &c. In the polypi and polyzoa may be observed those resemblances in appearance which induced early naturalists to group them together, and also the wide difference of organization which marks the higher rank to which the latter have attained. Amongst the ciliated infusoria important gradations and differences will also be noticed, some having only one sort of cilia, others two sorts, and others, again, supplied, in addition to cilia, with hooks and styles. No perfectly satisfactory classification of the infusoria has yet been devised, and the life history of a great many is still very imperfectly known. On the whole, the tendency of research is to place many of them higher than they used to stand after Ehrenberg's supposition of their having a plurality of distinct stomachs, &c., was given up. Balbiani and others have shown numerous cases of their forming their eggs by a process analogous to that of higher animals. Some really are, and others closely resemble, the larval conditions of creatures higher in the scale, and the contracted vesicle with its channel bears resemblance to what is called the "water vascular system" of worms.

Zoological classification depends very much on morphology, that is, the tracing of particular structures, or parts, through all their stages, from the lowest to the highest forms in which they are exhibited. In this way the swimming bladder of a fish is shown to be a rudimentary lung, though it has no respiratory functions, and Mr. Kitchen Parker has found in the imperfect skull of the tadpole a rudimentary appearance of bones belonging to the human ear. The comparative anatomist, after a wide survey of the objects before him, arranges them into groups. He asks what are the characteristic things to be affirmed concerning all the A's that cannot be said of all the B's; or of all the C's that marks their difference from the A's or the D's. Careful investigation upon these methods shows affinities where they were not previously expected—birds and reptiles being close relations, for example, instead of distant connections—and they lessen the value for purposes of classification of peculiarities that might have been deemed of the highest importance.

Professor Huxley divides the vertebrates into ITHYCOIDS, comprising fishes and amphibia, which, besides other characteristics, have gills at some period of their existence; SAUROIDS (reptiles and birds), which have no gills, and possess certain developmental characteristics in common; and, lastly, MAMMALS. The Insecta, Myriopoda, Arachnidæ, and Crustacea, he remarks, "without doubt present so many characters in common as to form a very natural assemblage. All are provided with articulated limbs attached to a segmented body skeleton, the latter, like the skeleton of the limbs, being an 'exoskeleton,' or a bordering of that layer which corresponds with the outer part of the vertebrates. In others, at any rate in the embryonic condition, the nervous system is composed of a double chain of ganglia, united by longitudinal commissures, and the gullet passed between two of these commissures. No one of the members of these four classes is known to possess vibratile cilia. The great majority of these animals have a distinct heart, provided with valvular apertures, which are in communication with a peri-visceral cavity containing corpusculated blood." These four classes have constituted the larger group or "province" of *Articulata* or *Arthropoda.* Professor Huxley thinks that, notwithstanding "the marked differences" between the Annelida (worms) and the preceding Arthropods (joint-foots), their resemblances outweighing them—"the characters of the nervous system, and the frequently segmented body, with imperfect lateral appendages of the Annelida, necessitates their assemblage with the Arthropoda in one great division, or sub-kingdom, of ANNULOSA."

Tracing analogies between the Echinodermata (sea urchins, star-fish, &c.) and the Scolecida (intestinal worms), he places them together as *Annuloida.*

Cephalopoda, Pteropoda, Pulmo-gasteropoda, and Branchio-gasteropoda, having resemblances of nervous system, and "all possessing that remarkable buccal apparatus, the Odontophore," are placed together by him as ODONTOPHORA. The Odontophores (tooth-bearers) are familiar to microscopists as the so-called *palates* of mollusca. Placing with the above the lamellibranchial mollusks (mollusks with gills formed of lamellæ or little plates), Ascidioida (ascidians), Brachiopoda (lamp-sheds), and Polyzoa, in spite of their differences, he forms another great group, ANNULOIDA.

The Actinozoa (anemonies, &c.) and the Hydrozoa (polyps) constitute the CŒLENTERA of Frey and Leuckart. "In all these animals," says Professor Huxley, "the substance of the body is differentiated into those histological elements which have been termed cells, and the latter are previously disposed in two layers, one external and one internal, constituting the ectoderm and endoderm. Among animals which possess this histological structure the

Cœlenterata stand alone in having an alimentary canal, which is open at its inner end and communicates freely by this aperture with the general cavity of the body," and "all (unless the Ctenophora should prove a partial exception to the rule) are provided with very remarkable organs of offence or defence, called thread-cells or nematocysts." In describing the Polyps we have given illustrations of these weapons.

The remaining classes, which have been roughly associated as *Protozoa*, must evidently be rearranged. Sponges, Rhizopods (Amœbæ, &c.), and Gregarines, have strong resemblances, but recent researches may place the former higher. The Infusoria comprehend creatures too various to remain under one head, and very many of them too highly organized to be called "protozoons," or first life-forms.

Those who wish to pursue this subject further may consult Professor Huxley's 'Elements of Comparative Anatomy,' from which the preceding quotations have been taken.

A system of classification founded upon anatomical and developmental considerations frequently differs considerably from one we might arrive at if all the creatures were arranged according to the perfection of their faculties and the extent and accuracy of their relations to the external world. Such a classification would not in any way supersede the former, but it would prove very instructive and offer many valuable suggestions. Some years since, Professor Owen proposed to divide the Vertebrates according to the perfection of their brains, but other anatomists did not find his divisions sufficiently coincident with facts. Very little has been done towards an exact science of human phrenology. The difficulties remain pretty much as they were many years ago, and our comparative phrenology, if we may use such a term, is in a very imperfect state. When we come to the lower animals we do not know what peculiarities of the brain of an ant make it the recipient of a higher instinct, or give its possessor greater capacities for dealing with new and unexpected difficulties than are possessed by most other insects, and if any reader has a marine aquarium, and will make a few experiments in taming prawns, and watching their proceedings, he will discover symptoms of intelligence beyond what the structure of the creature would have led him to expect.

Animals usually possess some one leading characteristic to which their general structure is subordinated. Man stands alone in having the whole of his organization conformed to the demands of a thinking, ruling brain. To pass at once to the other extreme, we observe in the lower infusoria a restless locomotion, probably subservient to respiration, but utterly inconsistent with a well developed life of relation, or with manifestations of thought. The life of an animalcule may be summed up as a brief and restricted, but vigorous organic energy, and if the amount of change which a single creature can make in the external world, is inconceivably small, the labours of the entire race alter the conditions of a prodigious amount of matter. Microscopic vegetable life is an important agent in purifying water from the taint of decomposing organisms. By evolving oxygen it brings putrescent particles under the influence of a species of combustion, which, though slow, is as effectual as that which a furnace could accomplish. In this way minute moulds burn up decaying wood.

Microscopic animal life helps the regenerative process, and, together with the minute vegetable life, restores to the organic system myriads of tons of matter, which death and decay would have handed over to the inorganic world. In a very small pond or tank the quantity of this kind of work is soon appreciable, and if we reflect on the amazing amount of water all over the globe, including seas and oceans, which swarm with infusoria, the total effect produced in a single year must seem considerable, even when compared with that portion of the earth's crust that is subject to alteration from all other causes put together. If we add to the labour of the Infusoria those of other creatures whose organization can only be discovered by the microscope, and take in the foraminifera, polyps, polyzoa, &c., we shall have to record still larger obligations to minute forms of living things. The coral polyp builds reefs that constitute the chief characteristic of certain regions in the Pacific; foraminifera are forming or helping to form strata of considerable extent, while diatoms are making deposits many feet in thickness, composed of myriads of their silicious shells, or adding their contributions of silex, very large in the aggregate, to all sedimentary rocks. Testimony of this kind of work is found by the navigator who examines the ice in arctic seas, and it comes up with soundings from the ocean depths.

On the surface of the earth the amount of change produced is equally remarkable, although it leaves less permanent traces behind. As a rule no decomposition of organized matter takes place, no death of plants or animals, without infusorial life making its appearance, and disposing of no small portion of the spoil. Even in our climate the mass of matter thus annually affected is very large; but what must it not be in moist tropical lands, where every particle seems alive, and the race of life and death goes on at a speed, and to an extent scarcely conceivable by those who have not witnessed it.

Thus, if we look at the world of minute forms which the microscope reveals, there opens before us a spectacle of boundless extent. We see life manifested by the specks of jelly containing particles not aggregated into structure, and we see it gradually ascending in complexities of organization. In creatures whose habits and appearance seem most remote from our own, we find the elementary developments of the organs and powers that constitute our glory, and give us our power. Such studies assist us to conceive of the universe as a Cosmos, or Beautifully Organized Whole; and, although we cannot tell the object for which a single portion received its precise form, we trace everywhere relations of structure to means of existence and enjoyment, and are led to the conviction that all the actions and arrangements of the organic or inorganic worlds are due to a definite direction and co-ordination of a few simple forces, which implicitly and unerringly obey the dictates of an Omniscient Mind.

CPSIA information can be obtained
at www.ICGtesting.com
Printed in the USA
LVHW032108310323
743149LV00005B/772

9 789356 909311